T0235075

# Spon's Estimating Costs Guide to Roofing

## Other books by Bryan Spain

## also available from Spon Press

| | |
|---|---|
| *Spon's Estimating Costs Guide to Minor Landscaping, Gardening and External Works* (2005 edition) | Pb: 0-415-34410-7 |
| *Spon's Estimating Costs Guide to Finishings: painting and decorating, plastering and tiling* (2005 edition) | Pb: 0-415-34411-5 |
| *Spon's Estimating Costs Guide to Plumbing & Heating* (2004 edition) | Pb: 0-415-31855-6 |
| *Spon's Estimating Costs Guide to Electrical Works* (2004 edition) | Pb: 0-415-31853-X |
| *Spon's Estimating Costs Guide to Minor Works, Alterations and Repairs to Fire, Flood, Gate and Theft Damage* (2004 edition) | Pb: 0-415-31854-8 |
| *Spon's House Improvement Price Book: house extensions, storm damage work, alterations, loft conversions and insulation* (2003 edition) | Pb: 0-415-30938-7 |
| *Spon's First Stage Estimating Price Book* (2000 edition) | Pb: 0-415-23436-0 |
| *Spon's Construction Resource Handbook* (1998 edition) | Hb: 0-419-23680-5 |

Information and ordering details

For price availability and ordering visit our website **www.sponpress.com**

Alternatively our books are available from all good bookshops.

# Spon's Estimating Costs Guide to Roofing

## Unit rates and project costs

Bryan Spain

Spon's Contractors' Handbooks

LONDON AND NEW YORK

First published 2005 by Taylor & Francis

Published 2021 by Routledge
2 Park Square, Milton Park, Abingdon, Oxon OX14 4RN
605 Third Avenue, New York, NY 10017

*Routledge is an imprint of the Taylor & Francis Group, an informa business*



*Publisher's Note*
This book has been prepared from camera-ready copy supplied by the author.

*British Library Cataloguing in Publication Data*
A catalogue record for this book is available from the British Library

*Library of Congress Cataloguing in Publication Data*
A catalog record for this book has been requested

ISBN 13: 978-0-415-34412-8 (pbk)

# Contents

## Marley Thaxton tiles



## Marley Heritage tiles



## Marley Plain tiles



## Marley Modern tiles



## Marley Ludlow Major tiles



## Marley Ludlow Plus tiles



## Marley Anglia Plus tiles

**Marley Mendip tiles**



**Marley Malvern tiles**



**Marley Double Roman tiles**



**Marley Bold Roll tiles**



**Marley Wessex tiles**



**Lafarge Norfolk tiles**



**Lafarge Renown tiles**

# Preface

This edition of Spon's Estimating Costs Guide to Roofing covers most of the type of roofing work encountered in the domestic construction market . It is intended to provide accurate cost data for small roofing firms to enable them to prepare estimates and quotations more quickly and more accurately.

This need for speed and acccuracy is vital for all contractors operating in the competitive domestic construction market. Most contractors have the skills necessary to carry out the work together with the capacity for dealing with the setbacks that are part of the normal construction process. But they rarely have enough time to complete the many tasks that must be carried out in order to trade profitably. This book aims to help roofing contractors by providing thousands of rates for roofing work and, if used sensibly, it can save them valuable time in the preparation of their bids.

I have received a great deal of support in the research necessary for this type of book and I am grateful to those individuals and firms who have provided the cost data and other information. In particular, I am indebted to Mark Loughrey of Youds, Ellison & Co., Chartered Accountants of Hoylake (tel: 0151-632 3298 or www.yesl.uk.com), who are specialists in advising small construction businesses. Their research for the information in the business section is based on tax legislation in force in December 2004.

Although every care has been taken in the preparation of the book, neither the Publisher nor I can accept any responsibility for the use of the information provided by any firm or individual. Finally, I would welcome any constructive criticism of the book's contents and suggestions that could be incorporated into future editions.

Bryan Spain
www.costofdiy.com
December 2004

# Introduction

This edition of Spon's Estimating Costs Guide to Roofing follows the layout, style and contents of previous editions. The contents of the book cover unit rates, project costs, repairs, tool and equipment hire, general advice on business matters and other information useful to those involved in the commissioning and construction of roofing work. The unit rates section presents analytical rates for work up to about £50,000 in value and the business section covers advice on starting and running a business together with information on taxation and VAT matters.

## Materials

In the domestic construction market, contractors are not usually able to purchase materials in large quantities and cannot benefit from the discounts available to larger contractors. An average of 10% to 15% discount has been allowed on normal trade prices.

## Labour

The hourly labour rates for craftsmen and general operatives are based upon the current wage awards promulgated by the Building and Allied Trades Joint Industrial Council (BATJIC). These are set at:

| | |
|---|---|
| Craftsman | £14.00 |
| General operative | £11.00 |

These rates include provision for NIC Employers' contribution, CITB levy, insurances, public and annual holidays, severance pay and tool allowances where appropriate.

## Headings

The following column headings have been used.

| Unit | Labour | Hours<br>£ | Materials<br>£ | O & P<br>£ | Total<br>£ |
|---|---|---|---|---|---|
| m2 | 0.20 | 2.80 | 2.10 | 0.74 | 5.64 |

## Unit

This column shows the unit of measurement for the item description:

| | |
|---|---|
| nr | number |
| m | linear metre |
| m2 | square metre |
| m3 | cubic metre. |

## Labour

In the example shown, 0.20 represents the estimated time estimated to carry out one square metre of the described item, i.e. 0.20 hours.

## Hours

The entry of £2.90 is calculated by multiplying the entry in the Labour column by the labour rate of £14.00.

## Materials

This column displays the cost of the materials required to carry out one square metre of the described item, i.e. £2.10.

## O & P (Overheads and profit)

This has been set at 15% and is deemed to cover head office and site overheads including:

- heating
- lighting
- rent
- rates
- telephones
- secretarial services
- insurances
- finance charges
- transport
- small tools
- ladders
- scaffolding etc.

## Total

This is the total of the Hours, Materials and Overheads and Profit columns.

## Generally

It should be noted that in the Project Costs section the horizontal and vertical totals may not always coincide due to rounding off.

## Contracting

Tradesmen and small contractors can act as main contractors (working for a client direct) or as a sub-contractor working for another contractor. Although a contract exists between a sub-contractor and a main contractor, there is no contractual link between a sub-contractor and an Employer.

In general terms this means that the sub-contractor cannot make any claims against the Employer direct and vice-versa. It also means that the sub-contractor should not accept any instructions from the Employer or his representative because this could be taken as establishing a privity of contract between the two parties,

A sub-contractor must be aware of his role in the programme because if he causes a contractor to overrun the completion date for the main contract he may become liable for the full amount of liquidated damages on the main contract plus the cost of damages that the contractor and other sub-contractors may have suffered.

A well-organised sub-contractor will keep a full set of daily site records, staffing levels, plant on site, weather charts and such like. It also cannot be over emphasised that any verbal instructions that the sub-contractor receives, should be confirmed immediately to the contractor in writing with the name of the person who issued them.

This procedure is extremely important because it may eventually save the sub-contractor considerable expense if someone tries to lay the blame for delays to the contract at his door. It is also important that instructions should only be taken from the contractor and he should be informed if another party attempts to do so.

## Contractor's discount

Most sub-contracts allow for a discount to the contractor of 2½ % from the sub-contractor's account. This means that the sub-contractor must add this discount to his prices by adding 1/39th to his net rates.

## Payment and retention

Payment is normally made on a monthly basis. The sub-contractor should submit his account to the contractor who then incorporates it into his own payment application and passes it on to the Architect or Employer's representative for certification of payment. When the sub-contractor receives his payment it will be reduced by 5% retention.

This money is held by the Employer and will be released in two parts. The first part, or moiety, is paid at the completion of work and, in the sub-contractor's case, this may be either when he has finished his work or when the contractor has completed the contract as a whole (known as practical completion) depending upon the contract conditions. The second part is released at the end of the defects liability period.

## Defects liability period

This is the period of time (normally 12 months) during which the sub-contractor is contractually bound to return to the job to rectify any mistakes or bad pieces of workmanship. This could either be twelve months from when he completes his work or twelve months from when the main contract is completed depending upon the wording of the sub-contract.

## Period for completion

Usually, a sub-contractor will be given a period of time in which he must complete the work and he must ensure that he has the capability to do the work within that period. Failure to meet the agreed completion date could have serious consequences.

Under certain circumstances, however, particularly with nominated sub-contracts, the sub-contractor may be requested to state the period of time he requires to do the work. If this is the case, then careful thought must be given to the time inserted. Too short a time may put him at financial risk but too long a time may prejudice the opportunity of winning the contract.

## Damages for non-completion

A clause is usually inserted within each sub-contract stating that the sub-contractor is liable for the financial losses that contractor suffers due to the sub-contractor's non-completion of work on time. This will include the amount of liquidated and ascertained damages contained within the main contract, together with the contractor's own direct losses and the direct losses of his other sub-contractors. As can be seen, the potential cost to the sub-contractor can be large so he must take care to expedite the work with due diligence to avoid incurring these costs.

**Variations**

All sub-contracts contain a clause allowing the sub-contract work to be varied without invalidating it. The sub-contractor will normally be paid any additional cost he incurs in carrying out variations.

**Insurances**

The sub-contractor is responsible for insuring against injury to persons or property and against loss of plant and materials. These insurances could be taken out for each individual job, although it is more common to take out blanket policies based on the turnover the firm has achieved in the previous year.

**Extensions of time**

The sub-contractor will normally be entitled to a longer period of time to complete the work if he is delayed or interrupted by reasons beyond his control (known as an extension of time). Most sub-contracts list the reasons and in some cases the sub-contractor may also be entitled to additional monies as well as an extension of time.

**Domestic sub-contracts**

In domestic sub-contracts the contractor would obtain competitive quotations from various sub-contractors of his own choice and these may be based on a bill of quantities, specification and drawings, or schedules of work. Accompanying the enquiry should also be a form of sub-contract that the sub-contractor will be required to complete.

There are several points that may affect costs and which the sub-contractor should bear in mind. These are: -

1. Whether the rates and prices are to include for any contractor's discount (normally expressed as plus 1/39$^{th}$ to allow 2½%).

2. Whether the contractor is to supply any labour or plant to assist the sub-contractor in either carrying out any of the work or in off-loading materials.

3. What facilities (if any) the contractor will provide for the sub-contractor such as mess rooms, welfare facilities, office accommodation and storage facilities.

4. Whether the contractor is to dispose of the sub-contractor's rubbish.

## Contracting

Often a sub-contractor will find himself working under a private contract, written or implied. This usually takes the form of working for a domestic householder or a small factory owner and the following procedures usually apply in this type of work.

## Estimate

The initial approach would usually come from a purchaser, e.g. 'How much will it cost to have my house re-roofed?.' At this stage, he may only want an approximate cost in order to see if he can afford to have the work carried out, (as opposed to a quotation – see later). Therefore, a brief description of the work to be carried out together with an approximate price will suffice.

However, it should be made clear that the price is an estimate and does not constitute an offer that may be accepted by the purchaser. The estimate may be based on a telephone conversation only, e.g. 'It will cost £8,000 to £ 10,000 to re-roof your house', or it could be based on a brief visit to the house. In either case, little time should be spent on an estimate and it is generally wise to express it as a price range.

## Quotation

A quotation is generally seen as an offer to do the work for the price quoted, and could constitute a simple contract if accepted. It follows that some time and effort should be spent in compiling a quotation to save arguments at a later stage. One should always remember that the roofer is the expert and must use his expertise in order to guide the purchaser and should discuss the work with him in full. He should tell the purchaser exactly what he is getting for the price and also what he is not.

This may mean going in to some detail such as what will happen to the old roof coverings, how access will be gained, how long the job will take, whether tarpaulins will be provided.

It is also recommended that any possible extras are also listed and priced separately at the time of quotation, e.g. renewal of chimney flashings. By the same token, the roofer should find out from the purchaser exactly what work he will do for the roofer and what restrictions (if any) he will place upon him.

For instance, will the purchaser keep the drive clear of cars to allow a skip and hoist to be used and will the roofer only be allowed entry to the premises on certain days and/or at certain times? These factors, should be ascertained in advance, and the costs of complying with them should be made known to the purchaser who may decide to take steps to change the restrictions.

Once the roofer has considered all the relevant factors then the formal written quotation can be produced. It should state precisely what the purchaser is getting for his money, including when and how long the job will take and contain all the salient points of discussions that have taken place.

After a quotation has been submitted then all that needs to be done is for the purchaser to accept it. Although a verbal acceptance would constitute a binding agreement, it is always more satisfactory if the acceptance is made in writing.

## Payments

There is much debate on how and when payments should be made in domestic situations. Ideally from a roofer's point of view to be paid in advance would be the most advantageous, but the chances of the purchaser wishing to do this are remote.

On the other hand, it may cause undue financial hardship on a recently self-employed roofer to have to buy all the materials himself and not get paid until all the work is completed. Whatever payment policy is adopted it must be agreed with the purchaser in advance and form part of the written quotation.

Possible alternatives are: -

1. Being paid when the work is complete. This is probably the best method from a public relations aspect and roofers who may be able to complete jobs in two or three days should have no difficulty in adopting this policy.

2. Being paid before the work is done. This is only really feasible where the roofer concerned is of unquestionable reputation or is well known to the purchaser.

3. Being paid for materials as they are bought and delivered and the balance paid when the work is complete. This could be a practical solution for smaller roofers, but the purchaser will probably want proof of the material costs, so careful handling of invoices is necessary.

4. Some form of stage payments usually take the form of agreed percentages of the quotation price or agreed parts of the quotation price paid after stages of the work have been carried out.

## Pricing and variations

It is important that some method of recording, pricing and being paid for variations is agreed at the outset and this is particularly relevant when dealing with private clients. Unforeseen additions, more than any other item, are the main cause of disputes and are often avoidable.

The first step to avoid this is to ensure that the original quotation is as detailed as possible. It is unhelpful for a quotation to read "Re-roof house £9,500'. If no drawings are involved, the roofer should list the main items of work he is quoting for with a price against each item as follows:

| | | £ |
|---|---|---|
| 1. | Take off all existing roof coverings and remove | 1,050.00 |
| 2. | New roof consisting of Marley 'Modern' tiles size 413 x 330mm on 19 x 38mm softwood battens including reinforced roofing felt | 3,765.00 |
| | Take off existing flashings to chimney and vent pipe and re-fix | 180.00 |
| | | 4,995 00 |
| | VAT @ 17.5% | 874.12 |
| | TOTAL QUOTATION | 5,869 12 |

The detailed specification of the materials could be contained within the descriptions or done separately. A quotation broken down in this way is detailed enough to enable the purchaser to ascertain that he is not being overcharged for any variations that may occur and yet is not so detailed that the purchaser is going to question the price of every detail.

Also, if the purchaser should wish to change anything himself then there are no arguments on what was included in the original quotation. If variations occur, it must be established who should pay for them. There are three main types of variations.

1. Those instructed by the purchaser.
2. Those that should have been included in the original quotation.
3. Those that are necessary due to events that could not have been foreseen.

The liabilities for 1 and 2 are relatively straightforward. If the purchaser says he wants Marley 'Wessex' tiles instead of Marley 'Modern', then he must bear the additional cost. Conversely, if the roofer forgot to include the cost of the roofing felt in his quotation then it is only fair that he bears the cost.

Item 3 is more difficult. If it is the purchaser who is receiving the benefit of the variations and if they were not foreseeable, then it would be logical to assume that it is the purchaser who should bear the cost. One example would be if a roofer was replacing some roof coverings, that revealed rotting timbers underneath that required replacement.

Other instances may not be as clear cut as this example and it may become necessary to arrive at a cost sharing arrangement if genuine doubt exists. Variations should preferably be agreed in advance before the work is carried out. They should be recorded and signed by both parties and, wherever possible, priced in detail and agreed.

# Standard Method of Measurement/trades link

The contents of this book are presented under trade headings and the following table provides a link to the Standard Method of Measurement (SMM7).

**Roofing**

H60 Clay/concrete roof tiling
H61 Fibre cement slating
H62 Natural slating
H63 Reconstructed stone slating
H71 Lead sheet coverings
H72 Aluminium sheet coverings
H73 Copper sheet coverings
H74 Zinc sheet coverings
R10 Rainwater pipes and gutters

# Part One

## UNIT RATES

Woodwool slab decking

Fibre cement cladding

Galvanised steel sheeting

Translucent sheeting

Underfelt and battens

Clay/concrete roof tiling

Fibre cement slating

Natural slating

Reconstructed stone slating

Timber shingling

Lead sheet coverings

Copper sheet coverings

Aluminium sheet coverings

Zinc coverings

Built-up roofing

Rainwater pipes

Rainwater gutters

Roof outlets

Roof lights

| | Unit | Labour | Hours £ | Mat'ls £ | O & P £ | Total £ |
|---|---|---|---|---|---|---|
| **WOODWOOL SLAB DECKING** | | | | | | |
| Woodwool reinforced wood wool slabs in standard lengths fixed to timber joists | | | | | | |
| 50mm thick | | | | | | |
| 1800mm lengths | m2 | 0.68 | 9.52 | 17.75 | 4.09 | 31.36 |
| 2000mm lengths | m2 | 0.68 | 9.52 | 18.25 | 4.17 | 31.94 |
| 2400mm lengths | m2 | 0.68 | 9.52 | 18.75 | 4.24 | 32.51 |
| 2700mm lengths | m2 | 0.68 | 9.52 | 19.25 | 4.32 | 33.09 |
| 3000mm lengths | m2 | 0.68 | 9.52 | 19.75 | 4.39 | 33.66 |
| 75mm thick | | | | | | |
| 1800mm lengths | m2 | 0.74 | 10.36 | 27.80 | 5.72 | 43.88 |
| 2000mm lengths | m2 | 0.74 | 10.36 | 28.30 | 5.80 | 44.46 |
| 2400mm lengths | m2 | 0.74 | 10.36 | 28.80 | 5.87 | 45.03 |
| 2700mm lengths | m2 | 0.74 | 10.36 | 29.30 | 5.95 | 45.61 |
| 3000mm lengths | m2 | 0.74 | 10.36 | 29.80 | 6.02 | 46.18 |
| 100mm thick | | | | | | |
| 3000mm lengths | m2 | 0.80 | 11.20 | 36.00 | 7.08 | 54.28 |
| 3300mm lengths | m2 | 0.80 | 11.20 | 36.50 | 7.16 | 54.86 |
| 3600mm lengths | m2 | 0.80 | 11.20 | 37.00 | 7.23 | 55.43 |
| 125mm thick | | | | | | |
| 2400mm lengths | m2 | 0.86 | 12.04 | 39.40 | 7.72 | 59.16 |
| 2700mm lengths | m2 | 0.86 | 12.04 | 39.90 | 7.79 | 59.73 |
| 3000mm lengths | m2 | 0.86 | 12.04 | 40.40 | 7.87 | 60.31 |
| Extra for pre-screeded | | | | | | |
| pre-screeded deck | m2 | - | - | 1.35 | 0.20 | 1.55 |
| pre-screeded soffit | m2 | - | - | 3.24 | 0.49 | 3.73 |
| proofed deck | m2 | - | - | 4.15 | 0.62 | 4.77 |

| | Unit | Labour | Hours £ | Mat'ls £ | O & P £ | Total £ |
|---|---|---|---|---|---|---|
| **FIBRE CEMENT CLADDING** | | | | | | |
| Corrugated fibre cement sheet cladding with one corrugation side lap and 150mm top and bottom laps fixed to sloping surfaces | | | | | | |
| profile 3, grey | | | | | | |
| fixed to timber purlins | m2 | 0.32 | 4.48 | 14.82 | 2.90 | 22.20 |
| fixed to steel purlins | m2 | 0.36 | 5.04 | 15.11 | 3.02 | 23.17 |
| profile 3, coloured | | | | | | |
| fixed to timber purlins | m2 | 0.32 | 4.48 | 16.56 | 3.16 | 24.20 |
| fixed to steel purlins | m2 | 0.36 | 5.04 | 16.89 | 3.29 | 25.22 |
| profile 6, grey | | | | | | |
| fixed to timber purlins | m2 | 0.36 | 5.04 | 15.32 | 3.05 | 23.41 |
| fixed to steel purlins | m2 | 0.40 | 5.60 | 15.65 | 3.19 | 24.44 |
| profile 6, coloured | | | | | | |
| fixed to timber purlins | m2 | 0.36 | 5.04 | 15.80 | 3.13 | 23.97 |
| fixed to steel purlins | m2 | 0.40 | 5.60 | 16.23 | 3.27 | 25.10 |
| Corrugated fibre cement sheet cladding with one corrugation side lap and 150mm top and bottom laps fixed to vertical surfaces | | | | | | |
| profile 3, grey | | | | | | |
| fixed to timber purlins | m2 | 0.38 | 5.32 | 14.82 | 3.02 | 23.16 |
| fixed to steel purlins | m2 | 0.42 | 5.88 | 15.11 | 3.15 | 24.14 |

| | Unit | Labour | Hours £ | Mat'ls £ | O & P £ | Total £ |
|---|---|---|---|---|---|---|
| profile 3, coloured | | | | | | |
| fixed to timber purlins | m2 | 0.38 | 5.32 | 16.56 | 3.28 | 25.16 |
| fixed to steel purlins | m2 | 0.42 | 5.88 | 16.89 | 3.42 | 26.19 |
| profile 6, grey | | | | | | |
| fixed to timber purlins | m2 | 0.42 | 5.88 | 15.32 | 3.18 | 24.38 |
| fixed to steel purlins | m2 | 0.46 | 6.44 | 15.65 | 3.31 | 25.40 |
| profile 6, coloured | | | | | | |
| fixed to timber purlins | m2 | 0.42 | 5.88 | 15.80 | 3.25 | 24.93 |
| fixed to steel purlins | m2 | 0.46 | 6.44 | 16.23 | 3.40 | 26.07 |
| Extra for profile 3, grey accessories | | | | | | |
| eaves filler | m | 0.22 | 3.08 | 12.12 | 2.28 | 17.48 |
| apron flashing | m | 0.28 | 3.92 | 12.44 | 2.45 | 18.81 |
| ridge capping | m | 0.30 | 4.20 | 12.00 | 2.43 | 18.63 |
| ventilating capping | m | 0.30 | 4.20 | 14.23 | 2.76 | 21.19 |
| vertical closure | m | 0.16 | 2.24 | 12.12 | 2.15 | 16.51 |
| Extra for profile 3, coloured accessories | | | | | | |
| eaves filler | m | 0.22 | 3.08 | 14.56 | 2.65 | 20.29 |
| apron flashing | m | 0.28 | 3.92 | 14.88 | 2.82 | 21.62 |
| ridge capping | m | 0.30 | 4.20 | 12.34 | 2.48 | 19.02 |
| ventilating capping | m | 0.30 | 4.20 | 17.95 | 3.32 | 25.47 |
| vertical closure | m | 0.16 | 2.24 | 14.56 | 2.52 | 19.32 |
| Extra for profile 6, grey accessories | | | | | | |
| eaves filler | m | 0.22 | 3.08 | 10.23 | 2.00 | 15.31 |
| apron flashing | m | 0.28 | 3.92 | 10.44 | 2.15 | 16.51 |
| ridge capping | m | 0.30 | 4.20 | 9.98 | 2.13 | 16.31 |
| ventilating capping | m | 0.30 | 4.20 | 19.55 | 3.56 | 27.31 |
| underglazing capping | m | 0.30 | 4.20 | 11.12 | 2.30 | 17.62 |
| vertical closure | m | 0.16 | 2.24 | 10.23 | 1.87 | 14.34 |

| | Unit | Labour | Hours £ | Mat'ls £ | O & P £ | Total £ |
|---|---|---|---|---|---|---|
| **Corrugated fibre cement sheet cladding (cont'd)** | | | | | | |
| Extra for profile 6, grey accessories | | | | | | |
| eaves filler | m | 0.22 | 3.08 | 12.44 | 2.33 | 17.85 |
| apron flashing | m | 0.28 | 3.92 | 12.65 | 2.49 | 19.06 |
| ridge capping | m | 0.30 | 4.20 | 12.22 | 2.46 | 18.88 |
| ventilating capping | m | 0.30 | 4.20 | 13.34 | 2.63 | 20.17 |
| underglazing capping | m | 0.30 | 4.20 | 11.12 | 2.30 | 17.62 |
| vertical closure | m | 0.16 | 2.24 | 12.44 | 2.20 | 16.88 |
| **GALVANISED STEEL SHEETING** | | | | | | |
| Galvanised steel strip troughed sheeting with standard light grey internal coat and Plastisol external coat, to sloping surfaces and fixed to steel purlins with self-tapping screws | | | | | | |
| 7mm thick with 19mm corrugations | m2 | 0.36 | 5.04 | 10.74 | 2.37 | 18.15 |
| 7mm thick with 32mm corrugations | m2 | 0.40 | 5.60 | 10.88 | 2.47 | 18.95 |
| 7mm thick with 38mm corrugations | m2 | 0.44 | 6.16 | 11.00 | 2.57 | 19.73 |
| 7mm thick with 46mm corrugations | m2 | 0.48 | 6.72 | 11.20 | 2.69 | 20.61 |
| 7mm thick with 60mm corrugations | m2 | 0.52 | 7.28 | 11.45 | 2.81 | 21.54 |

| | Unit | Labour | Hours £ | Mat'ls £ | O & P £ | Total £ |
|---|---|---|---|---|---|---|
| Galvanised steel strip troughed sheeting with standard light grey internal coat and Plastisol external coat, to vertical surfaces and fixed to steel rails with self-tapping screws | | | | | | |
| 7mm thick with 19mm corrugations | m2 | 0.40 | 5.60 | 10.74 | 2.45 | 18.79 |
| 7mm thick with 32mm corrugations | m2 | 0.44 | 6.16 | 10.88 | 2.56 | 19.60 |
| 7mm thick with 38mm corrugations | m2 | 0.48 | 6.72 | 11.00 | 2.66 | 20.38 |
| 7mm thick with 46mm corrugations | m2 | 0.52 | 7.28 | 11.20 | 2.77 | 21.25 |
| 7mm thick with 60mm corrugations | m2 | 0.56 | 7.84 | 11.45 | 2.89 | 22.18 |

**TRANSLUCENT SHEETING**

| | Unit | Labour | Hours £ | Mat'ls £ | O & P £ | Total £ |
|---|---|---|---|---|---|---|
| Glass fibre reinforced translucent sheeting with with one corrugation side lap and 150mm top and bottom laps fixed to sloping surfaces | | | | | | |
| profile 3 sheets | | | | | | |
| to timber members | m2 | 0.32 | 4.48 | 15.24 | 2.96 | 22.68 |
| to steel members | m2 | 0.36 | 5.04 | 15.45 | 3.07 | 23.56 |
| profile 6 sheets | | | | | | |
| to timber members | m2 | 0.36 | 5.04 | 15.24 | 3.04 | 23.32 |
| to steel members | m2 | 0.40 | 5.60 | 15.45 | 3.16 | 24.21 |

| | Unit | Labour | Hours £ | Mat'ls £ | O & P £ | Total £ |
|---|---|---|---|---|---|---|
| **Translucent sheeting (cont'd)** | | | | | | |
| Glass fibre reinforced translucent sheeting with with one corrugation side lap and 150mm top and bottom laps fixed to sloping surfaces | | | | | | |
| profile 3 sheets | | | | | | |
| to timber members | m2 | 0.36 | 5.04 | 15.24 | 3.04 | 23.32 |
| to steel members | m2 | 0.40 | 5.60 | 15.45 | 3.16 | 24.21 |
| profile 6 sheets | | | | | | |
| to timber members | m2 | 0.40 | 5.60 | 15.24 | 3.13 | 23.97 |
| to steel members | m2 | 0.44 | 6.16 | 15.45 | 3.24 | 24.85 |
| **UNDERFELT AND BATTENS** | | | | | | |
| Slaters' felt secured with battens | m2 | 0.03 | 0.42 | 1.25 | 0.25 | 1.92 |
| Slaters' reinforced felt secured with battens | m2 | 0.03 | 0.42 | 1.98 | 0.36 | 2.76 |
| Treated softwood counter-battens fixed to softwood joists with galvanised nails | | | | | | |
| 38 x 19mm | | | | | | |
| 450mm centres | m2 | 0.07 | 0.98 | 0.82 | 0.27 | 2.07 |
| 600mm centres | m2 | 0.05 | 0.70 | 0.66 | 0.20 | 1.56 |
| 750mm centres | m2 | 0.03 | 0.42 | 0.48 | 0.14 | 1.04 |

| | Unit | Labour | Hours £ | Mat'ls £ | O & P £ | Total £ |
|---|---|---|---|---|---|---|
| 38 x 25mm | | | | | | |
| 450mm centres | m2 | 0.08 | 1.12 | 0.98 | 0.32 | 2.42 |
| 600mm centres | m2 | 0.06 | 0.84 | 0.82 | 0.25 | 1.91 |
| 750mm centres | m2 | 0.04 | 0.56 | 0.64 | 0.18 | 1.38 |
| 38 x 38mm | | | | | | |
| 450mm centres | m2 | 0.10 | 1.40 | 1.24 | 0.40 | 3.04 |
| 600mm centres | m2 | 0.08 | 1.12 | 1.04 | 0.32 | 2.48 |
| 750mm centres | m2 | 0.06 | 0.84 | 0.83 | 0.25 | 1.92 |
| 46 x 38mm | | | | | | |
| 450mm centres | m2 | 0.12 | 1.68 | 1.42 | 0.47 | 3.57 |
| 600mm centres | m2 | 0.10 | 1.40 | 1.18 | 0.39 | 2.97 |
| 750mm centres | m2 | 0.08 | 1.12 | 1.02 | 0.32 | 2.46 |

## CLAY/CONCRETE ROOF TILING

| | Unit | Labour | Hours £ | Mat'ls £ | O & P £ | Total £ |
|---|---|---|---|---|---|---|
| Marley Marlden Plain granuled or smooth finish clay tiles size 267 x 168mm, 65mm lap, 35° pitch on reinforced underlay, gauge 100mm, battens size 38 × 25mm | m2 | 1.44 | 20.16 | 33.76 | 8.09 | 62.01 |
| Extra for | | | | | | |
| nailing every tile with aluminium nails | m2 | 0.15 | 2.10 | 2.40 | 0.68 | 5.18 |
| cloak verge system | m | 0.25 | 3.50 | 10.95 | 2.17 | 16.62 |
| double course at eaves | m | 0.90 | 12.60 | 3.10 | 2.36 | 18.06 |
| segmental ridge tile | m | 0.60 | 8.40 | 13.88 | 3.34 | 25.62 |
| valley tiles | m | 0.60 | 8.40 | 11.55 | 2.99 | 22.94 |
| bonnet hip tiles | m | 0.60 | 8.40 | 7.24 | 2.35 | 17.99 |
| eaves vent system | m | 0.60 | 8.40 | 8.05 | 2.47 | 18.92 |
| ventilated ridge terminal | nr | 0.60 | 8.40 | 37.65 | 6.91 | 52.96 |

| | Unit | Labour | Hours £ | Mat'ls £ | O & P £ | Total £ |
|---|---|---|---|---|---|---|
| **Marley Marlden Plain tiles (cont'd)** | | | | | | |
| vent terminal | nr | 0.60 | 8.40 | 56.56 | 9.74 | 74.70 |
| straight cutting | m | 0.20 | 2.80 | 0.00 | 0.42 | 3.22 |
| curved cutting | m | 0.30 | 4.20 | 0.00 | 0.63 | 4.83 |
| holes for pipes | nr | 0.40 | 5.60 | 0.00 | 0.84 | 6.44 |
| Marley Thaxton Plain granuled or smooth finish clay tiles size 270 x 168mm, 70mm lap, 35° pitch on reinforced underlay, gauge 100mm, battens size 38 × 25mm | m2 | 1.44 | 20.16 | 34.24 | 8.16 | 62.56 |
| Extra for | | | | | | |
| nailing every tile with aluminium nails | m2 | 0.15 | 2.10 | 2.40 | 0.68 | 5.18 |
| cloak verge system | m | 0.25 | 3.50 | 10.95 | 2.17 | 16.62 |
| double course at eaves | m | 0.90 | 12.60 | 3.10 | 2.36 | 18.06 |
| segmental ridge tile | m | 0.60 | 8.40 | 13.88 | 3.34 | 25.62 |
| valley tiles | m | 0.60 | 8.40 | 11.55 | 2.99 | 22.94 |
| bonnet hip tiles | m | 0.60 | 8.40 | 7.24 | 2.35 | 17.99 |
| eaves vent system | m | 0.60 | 8.40 | 8.05 | 2.47 | 18.92 |
| ventilated ridge terminal | nr | 0.60 | 8.40 | 37.65 | 6.91 | 52.96 |
| vent terminal | nr | 0.60 | 8.40 | 56.56 | 9.74 | 74.70 |
| straight cutting | m | 0.20 | 2.80 | 0.00 | 0.42 | 3.22 |
| curved cutting | m | 0.30 | 4.20 | 0.00 | 0.63 | 4.83 |
| holes for pipes | nr | 0.40 | 5.60 | 0.00 | 0.84 | 6.44 |

| | Unit | Labour | Hours £ | Mat'ls £ | O & P £ | Total £ |
|---|---|---|---|---|---|---|
| Marley Heritage Plain granuled or smooth finish clay tiles size 267 x 168mm, 65mm lap, 35° pitch on reinforced underlay, gauge 100mm, battens size 38 × 25mm | m2 | 1.44 | 20.16 | 34.50 | 8.20 | 62.86 |
| Extra for | | | | | | |
| nailing every tile with aluminium nails | m2 | 0.15 | 2.10 | 2.40 | 0.68 | 5.18 |
| cloak verge system | m | 0.25 | 3.50 | 10.95 | 2.17 | 16.62 |
| double course at eaves | m | 0.90 | 12.60 | 3.10 | 2.36 | 18.06 |
| segmental ridge tile | m | 0.60 | 8.40 | 13.88 | 3.34 | 25.62 |
| valley tiles | m | 0.60 | 8.40 | 11.55 | 2.99 | 22.94 |
| bonnet hip tiles | m | 0.60 | 8.40 | 7.24 | 2.35 | 17.99 |
| eaves vent system | m | 0.60 | 8.40 | 8.05 | 2.47 | 18.92 |
| ventilated ridge terminal | nr | 0.60 | 8.40 | 37.65 | 6.91 | 52.96 |
| vent terminal | nr | 0.60 | 8.40 | 56.56 | 9.74 | 74.70 |
| straight cutting | m | 0.20 | 2.80 | 0.00 | 0.42 | 3.22 |
| curved cutting | m | 0.30 | 4.20 | 0.00 | 0.63 | 4.83 |
| holes for pipes | nr | 0.40 | 5.60 | 0.00 | 0.84 | 6.44 |
| Marley Plain granuled or smooth finish clay tiles size 267 x 168mm, 65mm lap, 35° pitch on reinforced underlay, gauge 100mm, battens size 38 × 25mm | m2 | 1.44 | 20.16 | 33.40 | 8.03 | 61.59 |
| Extra for | | | | | | |
| nailing every tile with aluminium nails | m2 | 0.15 | 2.10 | 2.40 | 0.68 | 5.18 |
| cloak verge system | m | 0.25 | 3.50 | 10.95 | 2.17 | 16.62 |

| | Unit | Labour | Hours £ | Mat'ls £ | O & P £ | Total £ |
|---|---|---|---|---|---|---|
| **Marley Plain tiles (cont'd)** | | | | | | |
| double course at eaves | m | 0.90 | 12.60 | 3.10 | 2.36 | 18.06 |
| segmental ridge tile | m | 0.60 | 8.40 | 13.88 | 3.34 | 25.62 |
| valley tiles | m | 0.60 | 8.40 | 11.55 | 2.99 | 22.94 |
| bonnet hip tiles | m | 0.60 | 8.40 | 7.24 | 2.35 | 17.99 |
| eaves vent system | m | 0.60 | 8.40 | 8.05 | 2.47 | 18.92 |
| ventilated ridge terminal | nr | 0.60 | 8.40 | 37.65 | 6.91 | 52.96 |
| vent terminal | nr | 0.60 | 8.40 | 56.56 | 9.74 | 74.70 |
| straight cutting | m | 0.20 | 2.80 | 0.00 | 0.42 | 3.22 |
| curved cutting | m | 0.30 | 4.20 | 0.00 | 0.63 | 4.83 |
| holes for pipes | nr | 0.40 | 5.60 | 0.00 | 0.84 | 6.44 |
| Marley Modern smooth finish concrete interlocking tiles size 420 × 330mm, battens size 38 ×25mm, reinforced underlay, 75mm lap, 22.5 to 45° pitch battens size 38 × 25mm | m2 | 0.59 | 8.26 | 10.47 | 2.81 | 21.54 |
| Extra for | | | | | | |
| nailing every tile with aluminium nails | m2 | 0.02 | 0.28 | 0.40 | 0.10 | 0.78 |
| verge, cloak system | m | 0.50 | 7.00 | 3.93 | 1.64 | 12.57 |
| dry verge system | m | 0.60 | 8.40 | 4.86 | 1.99 | 15.25 |
| Modern ridge tile | m | 0.40 | 5.60 | 7.37 | 1.95 | 14.92 |
| Modern monoridge | m | 0.60 | 8.40 | 13.16 | 3.23 | 24.79 |
| dry ridge system | m | 0.60 | 8.40 | 12.94 | 3.20 | 24.54 |
| 1/3 hip riles | m | 0.40 | 5.60 | 7.24 | 1.93 | 14.77 |
| eaves vent system | m | 0.60 | 8.40 | 12.94 | 3.20 | 24.54 |
| ventilated ridge terminal | nr | 0.50 | 7.00 | 39.81 | 7.02 | 53.83 |
| gas vent terminal | nr | 0.50 | 7.00 | 56.56 | 9.53 | 73.09 |
| Marvent | nr | 0.50 | 7.00 | 19.27 | 3.94 | 30.21 |

| | Unit | Labour | Hours £ | Mat'ls £ | O & P £ | Total £ |
|---|---|---|---|---|---|---|
| straight cutting | m | 0.20 | 2.80 | 0.00 | 0.42 | 3.22 |
| holes for pipes | nr | 0.40 | 5.60 | 0.00 | 0.84 | 6.44 |
| Marley Ludlow Major concrete interlocking tiles size 420 × 330mm, battens size 38 ×25mm, reinforced underlay, 75mm lap, 22.5 to 45° pitch, battens size 38 × 25mm | m2 | 0.59 | 8.26 | 10.56 | 2.82 | 21.64 |
| Extra for | | | | | | |
| nailing every tile with aluminium nails | m2 | 0.02 | 0.28 | 0.40 | 0.10 | 0.78 |
| dry verge system | m | 0.60 | 8.40 | 4.86 | 1.99 | 15.25 |
| segmental ridge tile | m | 0.40 | 5.60 | 7.14 | 1.91 | 14.65 |
| segmental monoridge | m | 0.60 | 8.40 | 13.37 | 3.27 | 25.04 |
| dry ridge system | m | 0.60 | 8.40 | 12.94 | 3.20 | 24.54 |
| 1/3 hip riles | m | 0.40 | 5.60 | 7.24 | 1.93 | 14.77 |
| eaves vent system | m | 0.60 | 8.40 | 12.94 | 3.20 | 24.54 |
| ventilated ridge terminal | nr | 0.50 | 7.00 | 39.81 | 7.02 | 53.83 |
| gas vent terminal | nr | 0.50 | 7.00 | 56.56 | 9.53 | 73.09 |
| Marvent | nr | 0.50 | 7.00 | 19.27 | 3.94 | 30.21 |
| straight cutting | m | 0.20 | 2.80 | 0.00 | 0.42 | 3.22 |
| holes for pipes | nr | 0.40 | 5.60 | 0.00 | 0.84 | 6.44 |
| Marley Ludlow Plus concrete interlocking tiles size 387 × 229mm, battens size 38 ×25mm, reinforced underlay, 75mm lap, 22.5 to 45° pitch, battens size 38 × 25mm | m2 | 0.69 | 9.66 | 11.76 | 3.21 | 24.63 |

| | Unit | Labour | Hours £ | Mat'ls £ | O & P £ | Total £ |
|---|---|---|---|---|---|---|
| **Marley Ludlow Plus (cont'd)** | | | | | | |
| Extra for | | | | | | |
| nailing every tile with aluminium nails | m2 | 0.03 | 0.42 | 0.60 | 0.15 | 1.17 |
| segmental ridge tile | m | 0.40 | 5.60 | 7.14 | 1.91 | 14.65 |
| segmental monoridge | m | 0.60 | 8.40 | 13.37 | 3.27 | 25.04 |
| dry ridge system | m | 0.60 | 8.40 | 12.94 | 3.20 | 24.54 |
| 1/3 hip riles | m | 0.40 | 5.60 | 7.24 | 1.93 | 14.77 |
| eaves vent system | m | 0.60 | 8.40 | 12.94 | 3.20 | 24.54 |
| ventilated ridge terminal | nr | 0.50 | 7.00 | 39.81 | 7.02 | 53.83 |
| gas vent terminal | nr | 0.50 | 7.00 | 56.56 | 9.53 | 73.09 |
| Marvent | nr | 0.50 | 7.00 | 19.27 | 3.94 | 30.21 |
| straight cutting | m | 0.20 | 2.80 | 0.00 | 0.42 | 3.22 |
| holes for pipes | nr | 0.40 | 5.60 | 0.00 | 0.84 | 6.44 |
| Marley Anglia Plus concrete interlocking tiles size 387 × 229mm, battens size 38 ×25mm, reinforced underlay, 75mm lap, 22.5 to 45° pitch, battens size 38 × 25mm | m2 | 0.69 | 9.66 | 13.97 | 3.54 | 27.17 |
| Extra for | | | | | | |
| nailing every tile with aluminium nails | m2 | 0.03 | 0.42 | 0.60 | 0.15 | 1.17 |
| segmental ridge tile | m | 0.40 | 5.60 | 7.14 | 1.91 | 14.65 |
| segmental monoridge | m | 0.60 | 8.40 | 13.37 | 3.27 | 25.04 |
| dry ridge system | m | 0.60 | 8.40 | 12.94 | 3.20 | 24.54 |
| 1/3 hip riles | m | 0.40 | 5.60 | 7.24 | 1.93 | 14.77 |
| eaves vent system | m | 0.60 | 8.40 | 12.94 | 3.20 | 24.54 |

| | Unit | Labour | Hours £ | Mat'ls £ | O & P £ | Total £ |
|---|---|---|---|---|---|---|
| ventilated ridge terminal | nr | 0.50 | 7.00 | 39.81 | 7.02 | 53.83 |
| gas vent terminal | nr | 0.50 | 7.00 | 56.56 | 9.53 | 73.09 |
| Marvent | nr | 0.50 | 7.00 | 19.27 | 3.94 | 30.21 |
| straight cutting | m | 0.20 | 2.80 | 0.00 | 0.42 | 3.22 |
| holes for pipes | nr | 0.40 | 5.60 | 0.00 | 0.84 | 6.44 |
| Marley Mendip concrete interlocking tiles size 420 × 330mm, battens size 38 ×25mm, reinforced underlay, 75mm lap, 22.5 to 45° pitch, battens size 38 × 25mm | m2 | 0.59 | 8.26 | 10.74 | 2.85 | 21.85 |
| Extra for | | | | | | |
| nailing every tile with aluminium nails | m2 | 0.02 | 0.28 | 0.40 | 0.10 | 0.78 |
| dry verge system | m | 0.60 | 8.40 | 4.86 | 1.99 | 15.25 |
| verge, cloak system | m | 0.50 | 7.00 | 3.93 | 1.64 | 12.57 |
| 1/3 hip riles | m | 0.40 | 5.60 | 7.24 | 1.93 | 14.77 |
| segmental monoridge | m | 0.60 | 8.40 | 13.37 | 3.27 | 25.04 |
| dry ridge system | m | 0.60 | 8.40 | 12.94 | 3.20 | 24.54 |
| segmental hip tiles | m | 0.40 | 5.60 | 6.91 | 1.88 | 14.39 |
| eaves vent system | m | 0.60 | 8.40 | 12.94 | 3.20 | 24.54 |
| ventilated ridge terminal | nr | 0.50 | 7.00 | 39.81 | 7.02 | 53.83 |
| gas vent terminal | nr | 0.50 | 7.00 | 56.56 | 9.53 | 73.09 |
| Marvent | nr | 0.50 | 7.00 | 19.27 | 3.94 | 30.21 |
| straight cutting | m | 0.20 | 2.80 | 0.00 | 0.42 | 3.22 |
| holes for pipes | nr | 0.40 | 5.60 | 0.00 | 0.84 | 6.44 |

| | Unit | Labour | Hours £ | Mat'ls £ | O & P £ | Total £ |
|---|---|---|---|---|---|---|
| Marley Malvern concrete interlocking tiles size 420 × 330mm, battens size 38 ×25mm, reinforced underlay, 75mm lap, 22.5 to 45° pitch battens size 38 × 25mm | m2 | 0.59 | 8.26 | 10.35 | 2.79 | 21.40 |
| Extra for | | | | | | |
| nailing every tile with aluminium nails | m2 | 0.02 | 0.28 | 0.40 | 0.10 | 0.78 |
| dry verge system | m | 0.60 | 8.40 | 4.86 | 1.99 | 15.25 |
| segmental ridge tile | m | 0.40 | 5.60 | 7.14 | 1.91 | 14.65 |
| segmental monoridge | m | 0.60 | 8.40 | 13.37 | 3.27 | 25.04 |
| dry ridge system | m | 0.60 | 8.40 | 12.94 | 3.20 | 24.54 |
| 1/3 hip riles | m | 0.40 | 5.60 | 7.24 | 1.93 | 14.77 |
| eaves vent system | m | 0.60 | 8.40 | 12.94 | 3.20 | 24.54 |
| ventilated ridge terminal | nr | 0.50 | 7.00 | 39.81 | 7.02 | 53.83 |
| gas vent terminal | nr | 0.50 | 7.00 | 56.56 | 9.53 | 73.09 |
| Marvent | nr | 0.50 | 7.00 | 19.27 | 3.94 | 30.21 |
| straight cutting | m | 0.20 | 2.80 | 0.00 | 0.42 | 3.22 |
| holes for pipes | nr | 0.40 | 5.60 | 0.00 | 0.84 | 6.44 |
| segmental monoridge | m | 0.60 | 8.40 | 13.37 | 3.27 | 25.04 |
| dry ridge system | m | 0.60 | 8.40 | 12.94 | 3.20 | 24.54 |
| segmental hip tiles | m | 0.40 | 5.60 | 6.91 | 1.88 | 14.39 |
| eaves vent system | m | 0.60 | 8.40 | 12.94 | 3.20 | 24.54 |
| ventilated ridge terminal | nr | 0.50 | 7.00 | 39.81 | 7.02 | 53.83 |
| gas vent terminal | nr | 0.50 | 7.00 | 56.56 | 9.53 | 73.09 |
| Marvent | nr | 0.50 | 7.00 | 19.27 | 3.94 | 30.21 |
| straight cutting | m | 0.20 | 2.80 | 0.00 | 0.42 | 3.22 |
| holes for pipes | nr | 0.40 | 5.60 | 0.00 | 0.84 | 6.44 |

| | Unit | Labour | Hours £ | Mat'ls £ | O & P £ | Total £ |
|---|---|---|---|---|---|---|
| Marley Double Roman concrete interlocking tiles size 420 × 330mm, battens size 38 ×25mm, reinforced underlay, 75mm lap, 22.5 to 45° pitch battens size 38 × 25mm | m2 | 0.59 | 8.26 | 10.02 | 2.74 | 21.02 |
| Extra for | | | | | | |
| nailing every tile with aluminium nails | m2 | 0.02 | 0.28 | 0.40 | 0.10 | 0.78 |
| dry verge system | m | 0.60 | 8.40 | 4.86 | 1.99 | 15.25 |
| verge, cloak system | m | 0.50 | 7.00 | 3.93 | 1.64 | 12.57 |
| segmental ridge tile | m | 0.40 | 5.60 | 7.14 | 1.91 | 14.65 |
| segmental monoridge | m | 0.60 | 8.40 | 13.37 | 3.27 | 25.04 |
| dry ridge system | m | 0.60 | 8.40 | 12.94 | 3.20 | 24.54 |
| 1/3 hip riles | m | 0.40 | 5.60 | 7.24 | 1.93 | 14.77 |
| eaves vent system | m | 0.60 | 8.40 | 12.94 | 3.20 | 24.54 |
| ventilated ridge terminal | nr | 0.50 | 7.00 | 39.81 | 7.02 | 53.83 |
| gas vent terminal | nr | 0.50 | 7.00 | 56.56 | 9.53 | 73.09 |
| Marvent | nr | 0.50 | 7.00 | 19.27 | 3.94 | 30.21 |
| straight cutting | m | 0.20 | 2.80 | 0.00 | 0.42 | 3.22 |
| holes for pipes | nr | 0.40 | 5.60 | 0.00 | 0.84 | 6.44 |
| Marley Bold Roll concrete interlocking tiles size 420 × 330mm, battens size 38 ×25mm, reinforced underlay, 75mm lap, 22.5 to 45° pitch battens size 38 × 25mm | m2 | 0.59 | 8.26 | 10.02 | 2.74 | 21.02 |
| Extra for | | | | | | |
| nailing every tile with aluminium nails | m2 | 0.02 | 0.28 | 0.40 | 0.10 | 0.78 |

| | Unit | Labour | Hours £ | Mat'ls £ | O & P £ | Total £ |
|---|---|---|---|---|---|---|
| dry verge system | m | 0.60 | 8.40 | 4.86 | 1.99 | 15.25 |
| verge, cloak system | m | 0.50 | 7.00 | 3.93 | 1.64 | 12.57 |
| segmental ridge tile | m | 0.40 | 5.60 | 7.14 | 1.91 | 14.65 |
| segmental monoridge | m | 0.60 | 8.40 | 13.37 | 3.27 | 25.04 |
| dry ridge system | m | 0.60 | 8.40 | 12.94 | 3.20 | 24.54 |
| 1/3 hip riles | m | 0.40 | 5.60 | 7.24 | 1.93 | 14.77 |
| eaves vent system | m | 0.60 | 8.40 | 12.94 | 3.20 | 24.54 |
| ventilated ridge terminal | nr | 0.50 | 7.00 | 39.81 | 7.02 | 53.83 |
| gas vent terminal | nr | 0.50 | 7.00 | 56.56 | 9.53 | 73.09 |
| Marvent | nr | 0.50 | 7.00 | 19.27 | 3.94 | 30.21 |
| straight cutting | m | 0.20 | 2.80 | 0.00 | 0.42 | 3.22 |
| holes for pipes | nr | 0.40 | 5.60 | 0.00 | 0.84 | 6.44 |
| Marley Wessex concrete interlocking tiles size 420 × 330mm, battens size 38 ×25mm, reinforced underlay, 75mm lap, 22.5 to 45° pitch battens size 38 × 25mm | m2 | 0.59 | 8.26 | 14.41 | 3.40 | 26.07 |
| Extra for | | | | | | |
| nailing every tile with aluminium nails | m2 | 0.02 | 0.28 | 0.40 | 0.10 | 0.78 |
| dry verge system | m | 0.60 | 8.40 | 4.86 | 1.99 | 15.25 |
| segmental ridge tile | m | 0.40 | 5.60 | 7.14 | 1.91 | 14.65 |
| segmental monoridge | m | 0.60 | 8.40 | 13.37 | 3.27 | 25.04 |
| dry ridge system | m | 0.60 | 8.40 | 12.94 | 3.20 | 24.54 |
| 1/3 hip riles | m | 0.40 | 5.60 | 7.24 | 1.93 | 14.77 |
| eaves vent system | m | 0.60 | 8.40 | 12.94 | 3.20 | 24.54 |
| ventilated ridge terminal | nr | 0.50 | 7.00 | 39.81 | 7.02 | 53.83 |
| gas vent terminal | nr | 0.50 | 7.00 | 56.56 | 9.53 | 73.09 |
| Marvent | nr | 0.50 | 7.00 | 19.27 | 3.94 | 30.21 |
| straight cutting | m | 0.20 | 2.80 | 0.00 | 0.42 | 3.22 |
| holes for pipes | nr | 0.40 | 5.60 | 0.00 | 0.84 | 6.44 |

| | Unit | Labour | Hours £ | Mat'ls £ | O & P £ | Total £ |
|---|---|---|---|---|---|---|
| Lafarge Norfolk concrete interlocking pantiles size 381 × 227mm, battens size 38 ×25mm, reinforced underlay, 75mm lap, battens size 38 × 25mm | m2 | 0.69 | 9.66 | 12.58 | 3.34 | 25.58 |
| Extra for | | | | | | |
| nailing every tile with aluminium nails | m2 | 0.02 | 0.28 | 0.40 | 0.10 | 0.78 |
| extra single course at verge | m | 0.40 | 5.60 | 6.32 | 1.79 | 13.71 |
| cloaked verge course | m | 0.40 | 5.60 | 7.80 | 2.01 | 15.41 |
| eaves vent system | m | 0.40 | 5.60 | 9.87 | 2.32 | 17.79 |
| universal ridge tile | m | 0.40 | 5.60 | 9.99 | 2.34 | 17.93 |
| universal hip tile | m | 0.40 | 5.60 | 9.99 | 2.34 | 17.93 |
| valley trough tile | m | 0.40 | 5.60 | 8.11 | 2.06 | 15.77 |
| universal gas flue ridge tile | nr | 0.50 | 7.00 | 52.36 | 8.90 | 68.26 |
| straight cutting | m | 0.20 | 2.80 | 0.00 | 0.42 | 3.22 |
| holes for pipes | nr | 0.40 | 5.60 | 0.00 | 0.84 | 6.44 |
| Lafarge Renown concrete interlocking tiles size 418 × 330mm, battens size 38 ×25mm, reinforced underlay, 75mm lap, battens size 38 × 25mm | m2 | 0.59 | 8.26 | 10.46 | 2.81 | 21.53 |
| Extra for | | | | | | |
| nailing every tile with aluminium nails | m2 | 0.03 | 0.42 | 0.60 | 0.15 | 1.17 |
| extra single course at verge | m | 0.40 | 5.60 | 6.32 | 1.79 | 13.71 |
| cloaked verge course | m | 0.40 | 5.60 | 7.80 | 2.01 | 15.41 |
| eaves vent system | m | 0.40 | 5.60 | 9.87 | 2.32 | 17.79 |
| universal ridge tile | m | 0.40 | 5.60 | 9.99 | 2.34 | 17.93 |

| | Unit | Labour | Hours £ | Mat'ls £ | O & P £ | Total £ |
|---|---|---|---|---|---|---|
| **Lafarge Renown concrete tiles (cont'd)** | | | | | | |
| universal hip tile | m | 0.40 | 5.60 | 9.99 | 2.34 | 17.93 |
| valley trough tile | m | 0.40 | 5.60 | 8.11 | 2.06 | 15.77 |
| universal gas flue ridge tile | nr | 0.50 | 7.00 | 52.36 | 8.90 | 68.26 |
| straight cutting | m | 0.20 | 2.80 | 0.00 | 0.42 | 3.22 |
| holes for pipes | nr | 0.40 | 5.60 | 0.00 | 0.84 | 6.44 |
| Lafarge Double Roman concrete interlocking tiles size 418 × 330mm, battens size 38 × 25mm, reinforced underlay, 75mm lap battens size 38 × 25mm | m2 | 0.59 | 8.26 | 10.46 | 2.81 | 21.53 |
| Extra for | | | | | | |
| nailing every tile with aluminium nails | m2 | 0.03 | 0.42 | 0.60 | 0.15 | 1.17 |
| extra single course at verge | m | 0.40 | 5.60 | 6.32 | 1.79 | 13.71 |
| cloaked verge course | m | 0.40 | 5.60 | 7.80 | 2.01 | 15.41 |
| eaves vent system | m | 0.40 | 5.60 | 9.87 | 2.32 | 17.79 |
| universal ridge tile | m | 0.40 | 5.60 | 9.99 | 2.34 | 17.93 |
| universal hip tile | m | 0.40 | 5.60 | 9.99 | 2.34 | 17.93 |
| valley trough tile | m | 0.40 | 5.60 | 8.11 | 2.06 | 15.77 |
| universal gas flue ridge tile | nr | 0.50 | 7.00 | 52.36 | 8.90 | 68.26 |
| straight cutting | m | 0.20 | 2.80 | 0.00 | 0.42 | 3.22 |
| holes for pipes | nr | 0.40 | 5.60 | 0.00 | 0.84 | 6.44 |
| Lafarge Regent concrete interlocking tiles size 418 × 332mm, battens size 38 × 25mm, reinforced underlay, 75mm lap battens size 38 × 25mm | m2 | 0.59 | 8.26 | 10.75 | 2.85 | 21.86 |

| | Unit | Labour | Hours £ | Mat'ls £ | O & P £ | Total £ |
|---|---|---|---|---|---|---|
| Extra for | | | | | | |
| nailing every tile with aluminium nails | m2 | 0.03 | 0.42 | 0.60 | 0.15 | 1.17 |
| extra single course at verge | m | 0.40 | 5.60 | 6.32 | 1.79 | 13.71 |
| cloaked verge course | m | 0.40 | 5.60 | 7.80 | 2.01 | 15.41 |
| eaves vent system | m | 0.40 | 5.60 | 9.87 | 2.32 | 17.79 |
| universal ridge tile | m | 0.40 | 5.60 | 9.99 | 2.34 | 17.93 |
| universal hip tile | m | 0.40 | 5.60 | 9.99 | 2.34 | 17.93 |
| valley trough tile | m | 0.40 | 5.60 | 8.11 | 2.06 | 15.77 |
| universal gas flue ridge tile | nr | 0.50 | 7.00 | 52.36 | 8.90 | 68.26 |
| straight cutting | m | 0.20 | 2.80 | 0.00 | 0.42 | 3.22 |
| holes for pipes | nr | 0.40 | 5.60 | 0.00 | 0.84 | 6.44 |
| Lafarge Grovebury concrete interlocking tiles size 418 × 332mm, battens size 38 × 25mm, reinforced underlay, 75mm lap battens size 38 × 25mm | m2 | 0.59 | 8.26 | 10.75 | 2.85 | 21.86 |
| Extra for | | | | | | |
| nailing every tile with aluminium nails | m2 | 0.03 | 0.42 | 0.60 | 0.15 | 1.17 |
| extra single course at verge | m | 0.40 | 5.60 | 6.32 | 1.79 | 13.71 |
| cloaked verge course | m | 0.40 | 5.60 | 7.80 | 2.01 | 15.41 |
| eaves vent system | m | 0.40 | 5.60 | 9.87 | 2.32 | 17.79 |
| universal ridge tile | m | 0.40 | 5.60 | 9.99 | 2.34 | 17.93 |
| universal hip tile | m | 0.40 | 5.60 | 9.99 | 2.34 | 17.93 |
| valley trough tile | m | 0.40 | 5.60 | 8.11 | 2.06 | 15.77 |
| universal gas flue ridge tile | nr | 0.50 | 7.00 | 52.36 | 8.90 | 68.26 |
| straight cutting | m | 0.20 | 2.80 | 0.00 | 0.42 | 3.22 |
| holes for pipes | nr | 0.40 | 5.60 | 0.00 | 0.84 | 6.44 |

| | Unit | Labour | Hours £ | Mat'ls £ | O & P £ | Total £ |
|---|---|---|---|---|---|---|
| Lafarge Stonewold concrete interlocking tiles size 430 × 380mm, battens size 38 ×25mm, reinforced underlay, 75mm lap, battens size 38 × 25mm | m2 | 0.59 | 8.26 | 15.88 | 3.62 | 27.76 |
| Extra for | | | | | | |
| nailing every tile with aluminium nails | m2 | 0.03 | 0.42 | 0.60 | 0.15 | 1.17 |
| extra single course at verge | m | 0.40 | 5.60 | 6.32 | 1.79 | 13.71 |
| cloaked verge course | m | 0.40 | 5.60 | 7.80 | 2.01 | 15.41 |
| eaves vent system | m | 0.40 | 5.60 | 9.87 | 2.32 | 17.79 |
| universal ridge tile | m | 0.40 | 5.60 | 9.99 | 2.34 | 17.93 |
| universal hip tile | m | 0.40 | 5.60 | 9.99 | 2.34 | 17.93 |
| valley trough tile | m | 0.40 | 5.60 | 8.11 | 2.06 | 15.77 |
| universal gas flue ridge tile | nr | 0.50 | 7.00 | 52.36 | 8.90 | 68.26 |
| straight cutting | m | 0.20 | 2.80 | 0.00 | 0.42 | 3.22 |
| holes for pipes | nr | 0.40 | 5.60 | 0.00 | 0.84 | 6.44 |
| **FIBRE CEMENT SLATING** | | | | | | |
| Asbestos-free artificial slates 400 × 200mm, pitch 40-45°, size 38 × 25mm softwood battens, reinforced underlay | | | | | | |
| lap 90mm, gauge 155mm | m2 | 0.90 | 12.60 | 35.71 | 7.25 | 55.56 |

| | Unit | Labour | Hours £ | Mat'ls £ | O & P £ | Total £ |
|---|---|---|---|---|---|---|
| Asbestos-free artificial slates 500 × 250mm, pitch over 25°, size 38 × 25mm softwood battens, reinforced underlay | | | | | | |
| lap 90mm, gauge 205mm | m2 | 0.80 | 11.20 | 32.14 | 6.50 | 49.84 |
| Asbestos-free artificial slates 600 × 300mm, pitch over 25°, size 38 × 25mm softwood battens, reinforced underlay | | | | | | |
| lap 100mm, gauge 250mm | m2 | 0.70 | 9.80 | 28.47 | 5.74 | 44.01 |
| Extra for | | | | | | |
| verge, 150mm wide plain tile undercloak | m | 0.20 | 2.80 | 2.10 | 0.74 | 5.64 |
| double course at eaves | m | 0.35 | 4.90 | 6.60 | 1.73 | 13.23 |
| half-round ridge tile | m | 0.40 | 5.60 | 23.78 | 4.41 | 33.79 |
| valley tiles | m | 0.60 | 8.40 | 23.78 | 4.83 | 37.01 |
| straight cutting | m | 0.20 | 2.80 | 0.00 | 0.42 | 3.22 |
| holes for pipes | nr | 0.40 | 5.60 | 0.00 | 0.84 | 6.44 |

## NATURAL SLATING

| | Unit | Labour | Hours £ | Mat'ls £ | O & P £ | Total £ |
|---|---|---|---|---|---|---|
| Blue/grey Welsh slates size 405 × 205mm, 75mm lap, 50 × 25mm softwood battens, reinforced underlay | | | | | | |
| sloping | m2 | 1.20 | 16.80 | 52.88 | 10.45 | 80.13 |
| vertical | m2 | 1.40 | 19.60 | 52.88 | 10.87 | 83.35 |

| | Unit | Labour | Hours £ | Mat'ls £ | O & P £ | Total £ |
|---|---|---|---|---|---|---|
| Extra for | | | | | | |
| double eaves course | m | 0.50 | 7.00 | 24.78 | 4.77 | 36.55 |
| verge undercloak course | m | 0.70 | 9.80 | 18.55 | 4.25 | 32.60 |
| angled ridge or hip tiles | m | 0.70 | 9.80 | 16.09 | 3.88 | 29.77 |
| mitred hips including cutting both sides | m | 0.70 | 9.80 | 27.28 | 5.56 | 42.64 |
| straight cutting | m | 0.60 | 8.40 | 0.00 | 1.26 | 9.66 |
| holes for pipes | nr | 0.40 | 5.60 | 0.00 | 0.84 | 6.44 |
| fix only lead soakers | nr | 0.50 | 7.00 | 0.00 | 1.05 | 8.05 |
| Blue/grey Welsh slates size 510 × 255mm, 75mm lap, 50 × 25mm softwood battens, reinforced underlay | | | | | | |
| sloping | m2 | 1.00 | 14.00 | 55.11 | 10.37 | 79.48 |
| vertical | m2 | 1.20 | 16.80 | 55.11 | 10.79 | 82.70 |
| Extra for | | | | | | |
| double eaves course | m | 0.50 | 7.00 | 31.00 | 5.70 | 43.70 |
| verge undercloak course | m | 0.70 | 9.80 | 19.68 | 4.42 | 33.90 |
| angled ridge or hip tiles | m | 0.70 | 9.80 | 16.09 | 3.88 | 29.77 |
| mitred hips including cutting both sides | m | 0.70 | 9.80 | 27.28 | 5.56 | 42.64 |
| straight cutting | m | 0.60 | 8.40 | 0.00 | 1.26 | 9.66 |
| holes for pipes | nr | 0.40 | 5.60 | 0.00 | 0.84 | 6.44 |
| fix only lead soakers | nr | 0.50 | 7.00 | 0.00 | 1.05 | 8.05 |

| | Unit | Labour | Hours £ | Mat'ls £ | O & P £ | Total £ |
|---|---|---|---|---|---|---|
| Blue/grey Welsh slates size 610 × 305mm, 75mm lap, 50 × 25mm softwood battens, reinforced underlay | | | | | | |
| sloping | m2 | 0.80 | 11.20 | 58.67 | 10.48 | 80.35 |
| vertical | m2 | 1.00 | 14.00 | 58.67 | 10.90 | 83.57 |
| Extra for | | | | | | |
| double eaves course | m | 0.50 | 7.00 | 34.98 | 6.30 | 48.28 |
| verge undercloak course | m | 0.70 | 9.80 | 21.58 | 4.71 | 36.09 |
| angled ridge or hip tiles | m | 0.70 | 9.80 | 16.09 | 3.88 | 29.77 |
| mitred hips including cutting both sides | m | 0.70 | 9.80 | 27.28 | 5.56 | 42.64 |
| straight cutting | m | 0.60 | 8.40 | 0.00 | 1.26 | 9.66 |
| holes for pipes | nr | 0.40 | 5.60 | 0.00 | 0.84 | 6.44 |
| fix only lead soakers | nr | 0.50 | 7.00 | 0.00 | 1.05 | 8.05 |
| Westmorland green slates in random lengths 450 - 230mm, 75mm laps, 50 × 25mm softwood battens, reinforced underlay | | | | | | |
| sloping | m2 | 1.35 | 18.90 | 118.21 | 20.57 | 157.68 |
| vertical | m2 | 1.55 | 21.70 | 118.21 | 20.99 | 160.90 |
| Extra for | | | | | | |
| double eaves course | m | 0.50 | 7.00 | 34.98 | 6.30 | 48.28 |
| verge undercloak course | m | 0.70 | 9.80 | 21.58 | 4.71 | 36.09 |
| angled ridge or hip tiles | m | 0.70 | 9.80 | 16.09 | 3.88 | 29.77 |
| straight cutting | m | 0.60 | 8.40 | 0.00 | 1.26 | 9.66 |
| holes for pipes | nr | 0.40 | 5.60 | 0.00 | 0.84 | 6.44 |
| fix only lead soakers | nr | 0.50 | 7.00 | 0.00 | 1.05 | 8.05 |

| | Unit | Labour | Hours £ | Mat'ls £ | O & P £ | Total £ |
|---|---|---|---|---|---|---|
| **RECONSTRUCTED STONE SLATING** | | | | | | |
| Marley Monarch interlocking slate size 325 × 330mm, 38 × 25mm softwood battens, reinforced underlay | | | | | | |
| 75mm lap, pitch 25-90° | m2 | 0.60 | 8.40 | 26.74 | 5.27 | 40.41 |
| 100mm lap, pitch 25-90° | m2 | 0.65 | 9.10 | 26.74 | 5.38 | 41.22 |
| nailing every tile with aluminium nails | m2 | 0.02 | 0.28 | 0.40 | 0.10 | 0.78 |
| interlocking dry verge | m | 0.60 | 8.40 | 4.86 | 1.99 | 15.25 |
| eaves vent system | m | 0.60 | 8.40 | 12.94 | 3.20 | 24.54 |
| Modern ridge tiles | m | 0.40 | 5.60 | 7.37 | 1.95 | 14.92 |
| Modern hip tiles | m | 0.40 | 5.60 | 7.37 | 1.95 | 14.92 |
| segmental ridge tile | m | 0.40 | 5.60 | 7.14 | 1.91 | 14.65 |
| Modern monoridge | m | 0.60 | 8.40 | 13.16 | 3.23 | 24.79 |
| dry ridge system | m | 0.60 | 8.40 | 12.94 | 3.20 | 24.54 |
| segmental hip tiles | m | 0.40 | 5.60 | 6.91 | 1.88 | 14.39 |
| ventilated ridge terminal | nr | 0.50 | 7.00 | 39.81 | 7.02 | 53.83 |
| gas vent terminal | nr | 0.50 | 7.00 | 56.56 | 9.53 | 73.09 |
| straight cutting | m | 0.20 | 2.80 | 0.00 | 0.42 | 3.22 |
| holes for pipes | nr | 0.40 | 5.60 | 0.00 | 0.84 | 6.44 |
| Lafarge Cambrian through-coloured slates size 300 × 336mm, 38 × 25mm softwood battens, reinforced underlay | m2 | 0.60 | 8.40 | 25.36 | 5.06 | 38.82 |
| Extra for | | | | | | |
| nailing every tile with aluminium nails | m2 | 0.02 | 0.28 | 0.40 | 0.10 | 0.78 |

| | Unit | Labour | Hours £ | Mat'ls £ | O & P £ | Total £ |
|---|---|---|---|---|---|---|
| dryvent ridge | m | 0.75 | 10.50 | 16.66 | 4.07 | 31.23 |
| half-round ridge or hip tile | m | 0.40 | 5.60 | 9.99 | 2.34 | 17.93 |
| verge system | m | 0.40 | 5.60 | 11.02 | 2.49 | 19.11 |
| eaves vent system | m | 0.60 | 8.40 | 9.70 | 2.72 | 20.82 |
| dryvent ridge | m | 0.75 | 10.50 | 16.66 | 4.07 | 31.23 |
| half-round ridge or hip tile | m | 0.40 | 5.60 | 9.99 | 2.34 | 17.93 |
| gas flue ridge terminal | nr | 0.50 | 7.00 | 52.36 | 8.90 | 68.26 |
| valley trough tiles | m | 0.40 | 5.60 | 8.11 | 2.06 | 15.77 |
| straight cutting | m | 0.20 | 2.80 | 0.00 | 0.42 | 3.22 |
| holes for pipes | nr | 0.40 | 5.60 | 0.00 | 0.84 | 6.44 |

**TIMBER SHINGLING**

| | Unit | Labour | Hours £ | Mat'ls £ | O & P £ | Total £ |
|---|---|---|---|---|---|---|
| Sawn red cedar treated shingles size 25 x 38 x 400mm, 38 × 25mm softwood battens, reinforced underlay | m2 | 1.35 | 18.90 | 22.36 | 6.19 | 47.45 |
| Extra for | | | | | | |
| valleys, cut two sides | m | 0.28 | 3.92 | 4.22 | 1.22 | 9.36 |
| double eaves course | m | 0.28 | 3.92 | 4.22 | 1.22 | 9.36 |
| pre-formed ridge course | m | 0.56 | 7.84 | 5.68 | 2.03 | 15.55 |
| holes for pipes | nr | 0.20 | 2.80 | 0.00 | 0.42 | 3.22 |

| | Unit | Labour | Hours £ | Mat'ls £ | O & P £ | Total £ |
|---|---|---|---|---|---|---|
| **LEAD SHEET COVERINGS** | | | | | | |
| Milled lead roof coverings to roof with falls less than 10° | | | | | | |
| code 5 | m2 | 3.10 | 43.40 | 24.43 | 10.17 | 78.00 |
| code 6 | m2 | 3.20 | 44.80 | 26.98 | 10.77 | 82.55 |
| Milled lead roof coverings to roof with falls more than 10° but less than 50° | | | | | | |
| code 5 | m2 | 3.20 | 44.80 | 24.43 | 10.38 | 79.61 |
| code 6 | m2 | 3.30 | 46.20 | 26.98 | 10.98 | 84.16 |
| Milled lead roof coverings to roof with falls over 50° | | | | | | |
| code 5 | m2 | 3.30 | 46.20 | 24.43 | 10.59 | 81.22 |
| code 6 | m2 | 3.40 | 47.60 | 26.98 | 11.19 | 85.77 |
| Milled lead roof coverings to dormers | | | | | | |
| code 5 | m2 | 3.80 | 53.20 | 24.43 | 11.64 | 89.27 |
| code 6 | m2 | 4.00 | 56.00 | 26.98 | 12.45 | 95.43 |
| Lead flashings | | | | | | |
| code 5, horizontal | | | | | | |
| 150mm girth | m | 0.60 | 8.40 | 3.68 | 1.81 | 13.89 |
| 200mm girth | m | 0.70 | 9.80 | 5.86 | 2.35 | 18.01 |
| 300mm girth | m | 0.85 | 11.90 | 7.35 | 2.89 | 22.14 |

| | Unit | Labour | Hours £ | Mat'ls £ | O & P £ | Total £ |
|---|---|---|---|---|---|---|
| code 5, sloping | | | | | | |
| 150mm girth | m | 0.80 | 11.20 | 3.68 | 2.23 | 17.11 |
| 200mm girth | m | 0.85 | 11.90 | 5.86 | 2.66 | 20.42 |
| 300mm girth | m | 0.90 | 12.60 | 7.35 | 2.99 | 22.94 |
| Lead stepped flashings | | | | | | |
| code 5, sloping | | | | | | |
| 150mm girth | m | 0.90 | 12.60 | 4.18 | 2.52 | 19.30 |
| 200mm girth | m | 0.95 | 13.30 | 6.36 | 2.95 | 22.61 |
| 300mm girth | m | 1.00 | 14.00 | 7.85 | 3.28 | 25.13 |
| Lead aprons | | | | | | |
| code 5, horizontal | | | | | | |
| 200mm girth | m | 0.90 | 12.60 | 5.86 | 2.77 | 21.23 |
| 300mm girth | m | 0.95 | 13.30 | 7.35 | 3.10 | 23.75 |
| 400mm girth | m | 1.00 | 14.00 | 11.00 | 3.75 | 28.75 |
| code 5, sloping | | | | | | |
| 200mm girth | m | 0.95 | 13.30 | 5.86 | 2.87 | 22.03 |
| 300mm girth | m | 1.00 | 14.00 | 7.35 | 3.20 | 24.55 |
| 400mm girth | m | 1.05 | 14.70 | 11.00 | 3.86 | 29.56 |
| Lead sills | | | | | | |
| code 5, horizontal | | | | | | |
| 200mm girth | m | 0.80 | 11.20 | 5.86 | 2.56 | 19.62 |
| 300mm girth | m | 0.90 | 12.60 | 7.35 | 2.99 | 22.94 |
| 400mm girth | m | 1.00 | 14.00 | 11.00 | 3.75 | 28.75 |
| code 5, sloping | | | | | | |
| 200mm girth | m | 0.90 | 12.60 | 5.86 | 2.77 | 21.23 |
| 300mm girth | m | 1.00 | 14.00 | 7.35 | 3.20 | 24.55 |
| 400mm girth | m | 1.10 | 15.40 | 11.00 | 3.96 | 30.36 |

| | Unit | Labour | Hours £ | Mat'ls £ | O & P £ | Total £ |
|---|---|---|---|---|---|---|
| Lead hips | | | | | | |
| code 5, sloping | | | | | | |
| 150mm girth | m | 0.95 | 13.30 | 5.86 | 2.87 | 22.03 |
| 200mm girth | m | 1.15 | 16.10 | 7.35 | 3.52 | 26.97 |
| 300mm girth | m | 1.35 | 18.90 | 11.00 | 4.49 | 34.39 |
| Lead kerbs | | | | | | |
| code 5, sloping | | | | | | |
| 300mm girth | m | 1.00 | 14.00 | 7.35 | 3.20 | 24.55 |
| 400mm girth | m | 1.10 | 15.40 | 11.00 | 3.96 | 30.36 |
| Lead valleys | | | | | | |
| code 5, sloping | | | | | | |
| 400mm girth | m | 1.10 | 15.40 | 11.00 | 3.96 | 30.36 |
| 600mm girth | m | 1.25 | 17.50 | 14.66 | 4.82 | 36.98 |
| Lead gutters | | | | | | |
| code 5, sloping | | | | | | |
| 600mm girth | m | 1.25 | 17.50 | 14.66 | 4.82 | 36.98 |
| 800mm girth | m | 1.55 | 21.70 | 19.05 | 6.11 | 46.86 |
| Lead slates | | | | | | |
| code 5 | | | | | | |
| size 400 × 400mm with 200mm high collar 100mm diameter | nr | 1.50 | 21.00 | 10.05 | 4.66 | 35.71 |
| size 400 × 400mm with 200mm high collar 150mm diameter | nr | 1.70 | 23.80 | 10.12 | 5.09 | 39.01 |

| | Unit | Labour | Hours £ | Mat'ls £ | O & P £ | Total £ |
|---|---|---|---|---|---|---|
| Labours on leadwork | | | | | | |
| raking cutting | m | 0.30 | 4.20 | 0.00 | 0.63 | 4.83 |
| curved cutting | m | 0.40 | 5.60 | 0.00 | 0.84 | 6.44 |
| welted edge | m | 0.35 | 4.90 | 0.00 | 0.74 | 5.64 |
| beaded edge | m | 0.35 | 4.90 | 0.00 | 0.74 | 5.64 |
| wedging into groove | m | 0.30 | 4.20 | 0.00 | 0.63 | 4.83 |
| leadburned angle | m | 0.50 | 7.00 | 0.00 | 1.05 | 8.05 |
| leadburned seam | m | 0.50 | 7.00 | 0.00 | 1.05 | 8.05 |
| dressing over | | | | | | |
| glass and glazing bars | m | 0.50 | 7.00 | 0.00 | 1.05 | 8.05 |
| along roof corrugations | m | 0.80 | 11.20 | 0.00 | 1.68 | 12.88 |
| across roof corrugations | m | 1.10 | 15.40 | 0.00 | 2.31 | 17.71 |
| fillers 200mm girth | m | 0.50 | 7.00 | 0.00 | 1.05 | 8.05 |
| hollows 200mm girth | m | 0.50 | 7.00 | 0.00 | 1.05 | 8.05 |
| channels 200mm girth | m | 0.50 | 7.00 | 0.00 | 1.05 | 8.05 |

**COPPER SHEET COVERINGS**

| | Unit | Labour | Hours £ | Mat'ls £ | O & P £ | Total £ |
|---|---|---|---|---|---|---|
| Copper sheet coverings to roof with falls less than 10° | | | | | | |
| 0.55mm thick | m2 | 3.20 | 44.80 | 26.24 | 10.66 | 81.70 |
| 0.61mm thick | m2 | 3.20 | 44.80 | 27.64 | 10.87 | 83.31 |
| Copper sheet coverings to roof with falls more than 10° but less than 50° | | | | | | |
| 0.55mm thick | m2 | 3.50 | 49.00 | 26.24 | 11.29 | 86.53 |
| 0.61mm thick | m2 | 3.50 | 49.00 | 27.64 | 11.50 | 88.14 |

| | Unit | Labour | Hours £ | Mat'ls £ | O & P £ | Total £ |
|---|---|---|---|---|---|---|
| Copper sheet coverings to roof with falls more than 50° | | | | | | |
| 0.55mm thick | m2 | 4.00 | 56.00 | 26.24 | 12.34 | 94.58 |
| 0.61mm thick | m2 | 4.00 | 56.00 | 27.64 | 12.55 | 96.19 |
| Copper flashings | | | | | | |
| 0.55mm thick, horizontal | | | | | | |
| 150mm girth | m | 0.55 | 7.70 | 3.93 | 1.74 | 13.37 |
| 200mm girth | m | 0.65 | 9.10 | 5.25 | 2.15 | 16.50 |
| 300mm girth | m | 0.75 | 10.50 | 7.87 | 2.76 | 21.13 |
| 0.55mm thick, sloping | | | | | | |
| 150mm girth | m | 0.60 | 8.40 | 3.93 | 1.85 | 14.18 |
| 200mm girth | m | 0.70 | 9.80 | 5.25 | 2.26 | 17.31 |
| 300mm girth | m | 0.80 | 11.20 | 7.87 | 2.86 | 21.93 |
| 0.61mm thick, horizontal | | | | | | |
| 150mm girth | m | 0.55 | 7.70 | 4.15 | 1.78 | 13.63 |
| 200mm girth | m | 0.65 | 9.10 | 5.23 | 2.15 | 16.48 |
| 300mm girth | m | 0.75 | 10.50 | 8.29 | 2.82 | 21.61 |
| 0.61mm thick, sloping | | | | | | |
| 150mm girth | m | 0.60 | 8.40 | 4.15 | 1.88 | 14.43 |
| 200mm girth | m | 0.70 | 9.80 | 5.23 | 2.25 | 17.28 |
| 300mm girth | m | 0.80 | 11.20 | 8.29 | 2.92 | 22.41 |
| Copper stepped flashings | | | | | | |
| 0.55mm thick, sloping | | | | | | |
| 150mm girth | m | 0.70 | 9.80 | 4.43 | 2.13 | 16.36 |
| 200mm girth | m | 0.80 | 11.20 | 5.75 | 2.54 | 19.49 |
| 300mm girth | m | 0.90 | 12.60 | 8.37 | 3.15 | 24.12 |

| | Unit | Labour | Hours £ | Mat'ls £ | O & P £ | Total £ |
|---|---|---|---|---|---|---|
| 0.61mm thick, sloping | | | | | | |
| 150mm girth | m | 0.70 | 9.80 | 4.65 | 2.17 | 16.62 |
| 200mm girth | m | 0.80 | 11.20 | 7.73 | 2.84 | 21.77 |
| 300mm girth | m | 0.90 | 12.60 | 8.79 | 3.21 | 24.60 |
| Copper apron | | | | | | |
| 0.55mm thick | | | | | | |
| 200mm girth | m | 0.75 | 10.50 | 5.35 | 2.38 | 18.23 |
| 300mm girth | m | 0.80 | 11.20 | 8.02 | 2.88 | 22.10 |
| 400mm girth | m | 0.85 | 11.90 | 10.70 | 3.39 | 25.99 |
| 0.61mm thick | | | | | | |
| 200mm girth | m | 1.00 | 14.00 | 5.53 | 2.93 | 22.46 |
| 300mm girth | m | 1.05 | 14.70 | 8.29 | 3.45 | 26.44 |
| 400mm girth | m | 1.10 | 15.40 | 11.06 | 3.97 | 30.43 |
| Copper capping to ridge | | | | | | |
| 0.55mm thick | | | | | | |
| 200mm girth | m | 0.85 | 11.90 | 5.35 | 2.59 | 19.84 |
| 300mm girth | m | 0.90 | 12.60 | 8.02 | 3.09 | 23.71 |
| 400mm girth | m | 0.95 | 13.30 | 10.70 | 3.60 | 27.60 |
| 0.61mm thick | | | | | | |
| 200mm girth | m | 0.85 | 11.90 | 5.53 | 2.61 | 20.04 |
| 300mm girth | m | 0.90 | 12.60 | 8.29 | 3.13 | 24.02 |
| 400mm girth | m | 0.95 | 13.30 | 11.06 | 3.65 | 28.01 |
| Copper capping to hip | | | | | | |
| 0.55mm thick | | | | | | |
| 200mm girth | m | 0.85 | 11.90 | 5.35 | 2.59 | 19.84 |
| 300mm girth | m | 0.90 | 12.60 | 8.02 | 3.09 | 23.71 |
| 400mm girth | m | 0.95 | 13.30 | 10.70 | 3.60 | 27.60 |

| | Unit | Labour | Hours £ | Mat'ls £ | O & P £ | Total £ |
|---|---|---|---|---|---|---|
| **Copper capping (cont'd)** | | | | | | |
| 0.61mm thick | | | | | | |
| 200mm girth | m | 0.85 | 11.90 | 5.53 | 2.61 | 20.04 |
| 300mm girth | m | 0.90 | 12.60 | 8.29 | 3.13 | 24.02 |
| 400mm girth | m | 0.95 | 13.30 | 11.06 | 3.65 | 28.01 |
| Copper lining to valley | | | | | | |
| 0.55mm thick | | | | | | |
| 200mm girth | m | 0.85 | 11.90 | 5.35 | 2.59 | 19.84 |
| 300mm girth | m | 0.90 | 12.60 | 8.02 | 3.09 | 23.71 |
| 400mm girth | m | 0.95 | 13.30 | 10.70 | 3.60 | 27.60 |
| 0.61mm thick | | | | | | |
| 200mm girth | m | 0.85 | 11.90 | 5.53 | 2.61 | 20.04 |
| 300mm girth | m | 0.90 | 12.60 | 8.29 | 3.13 | 24.02 |
| 400mm girth | m | 0.95 | 13.30 | 11.06 | 3.65 | 28.01 |
| Copper lining to gutter | | | | | | |
| 0.55mm thick | | | | | | |
| 200mm girth | m | 0.85 | 11.90 | 5.35 | 2.59 | 19.84 |
| 300mm girth | m | 0.90 | 12.60 | 8.02 | 3.09 | 23.71 |
| 400mm girth | m | 0.95 | 13.30 | 10.70 | 3.60 | 27.60 |
| 0.61mm thick | | | | | | |
| 200mm girth | m | 0.85 | 11.90 | 5.53 | 2.61 | 20.04 |
| 300mm girth | m | 0.90 | 12.60 | 8.29 | 3.13 | 24.02 |
| 400mm girth | m | 0.95 | 13.30 | 11.06 | 3.65 | 28.01 |
| Copper soakers size | | | | | | |
| 175 × 175mm | nr | 0.25 | 3.50 | 1.72 | 0.78 | 6.00 |
| 300 × 175mm | nr | 0.25 | 3.50 | 2.04 | 0.83 | 6.37 |

| | Unit | Labour | Hours £ | Mat'ls £ | O & P £ | Total £ |
|---|---|---|---|---|---|---|
| Labours on copper | | | | | | |
| raking cutting | m | 0.15 | 2.10 | 0.00 | 0.32 | 2.42 |
| curved cutting | m | 0.20 | 2.80 | 0.00 | 0.42 | 3.22 |
| welted edge | m | 0.25 | 3.50 | 0.00 | 0.53 | 4.03 |
| beaded edge | m | 0.25 | 3.50 | 0.00 | 0.53 | 4.03 |
| wedging into groove | m | 0.25 | 3.50 | 0.00 | 0.53 | 4.03 |
| brazed angle | m | 1.00 | 14.00 | 0.00 | 2.10 | 16.10 |
| brazed seam | m | 1.00 | 14.00 | 0.00 | 2.10 | 16.10 |

## ALUMINIUM SHEET COVERINGS

| | Unit | Labour | Hours £ | Mat'ls £ | O & P £ | Total £ |
|---|---|---|---|---|---|---|
| Commercial grade sheet aluminium coverings to roof with falls less than 10° | | | | | | |
| 0.6mm thick | m2 | 3.00 | 42.00 | 7.48 | 7.42 | 56.90 |
| 0.9mm thick | m2 | 3.00 | 42.00 | 8.61 | 7.59 | 58.20 |
| Commercial grade sheet aluminium coverings to roof with falls more than 10° and less that 50° | | | | | | |
| 0.6mm thick | m2 | 3.30 | 46.20 | 7.48 | 8.05 | 61.73 |
| 0.9mm thick | m2 | 3.30 | 46.20 | 8.61 | 8.22 | 63.03 |
| Commercial grade sheet aluminium coverings to roof with falls more than 50° | | | | | | |
| 0.6mm thick | m2 | 3.80 | 53.20 | 7.48 | 9.10 | 69.78 |
| 0.9mm thick | m2 | 3.80 | 53.20 | 8.61 | 9.27 | 71.08 |

| | Unit | Labour | Hours £ | Mat'ls £ | O & P £ | Total £ |
|---|---|---|---|---|---|---|
| Aluminium flashings | | | | | | |
| 0.6mm thick, horizontal | | | | | | |
| 150mm girth | m | 0.55 | 7.70 | 1.12 | 1.32 | 10.14 |
| 200mm girth | m | 0.65 | 9.10 | 1.50 | 1.59 | 12.19 |
| 300mm girth | m | 0.75 | 10.50 | 2.24 | 1.91 | 14.65 |
| 0.6mm thick, sloping | | | | | | |
| 150mm girth | m | 0.60 | 8.40 | 1.12 | 1.43 | 10.95 |
| 200mm girth | m | 0.70 | 9.80 | 1.50 | 1.70 | 13.00 |
| 300mm girth | m | 0.80 | 11.20 | 2.24 | 2.02 | 15.46 |
| 0.9mm thick, horizontal | | | | | | |
| 150mm girth | m | 0.55 | 7.70 | 1.29 | 1.35 | 10.34 |
| 200mm girth | m | 0.65 | 9.10 | 1.72 | 1.62 | 12.44 |
| 300mm girth | m | 0.75 | 10.50 | 2.58 | 1.96 | 15.04 |
| 0.9mm thick, sloping | | | | | | |
| 150mm girth | m | 0.60 | 8.40 | 1.29 | 1.45 | 11.14 |
| 200mm girth | m | 0.70 | 9.80 | 1.72 | 1.73 | 13.25 |
| 300mm girth | m | 0.80 | 11.20 | 2.58 | 2.07 | 15.85 |
| Aluminium stepped flashings | | | | | | |
| 0.6mm thick, sloping | | | | | | |
| 150mm girth | m | 0.70 | 9.80 | 1.12 | 1.64 | 12.56 |
| 200mm girth | m | 0.80 | 11.20 | 1.50 | 1.91 | 14.61 |
| 300mm girth | m | 0.90 | 12.60 | 2.24 | 2.23 | 17.07 |
| 0.9mm thick, sloping | | | | | | |
| 150mm girth | m | 0.70 | 9.80 | 1.29 | 1.66 | 12.75 |
| 200mm girth | m | 0.80 | 11.20 | 1.72 | 1.94 | 14.86 |
| 300mm girth | m | 0.90 | 12.60 | 2.58 | 2.28 | 17.46 |

| | Unit | Labour | Hours £ | Mat'ls £ | O & P £ | Total £ |
|---|---|---|---|---|---|---|
| Aluminium apron | | | | | | |
| 0.6mm thick | | | | | | |
| 200mm girth | m | 0.75 | 10.50 | 1.50 | 1.80 | 13.80 |
| 300mm girth | m | 0.80 | 11.20 | 2.24 | 2.02 | 15.46 |
| 400mm girth | m | 0.85 | 11.90 | 2.99 | 2.23 | 17.12 |
| 0.9mm thick | | | | | | |
| 200mm girth | m | 1.00 | 14.00 | 1.72 | 2.36 | 18.08 |
| 300mm girth | m | 1.05 | 14.70 | 2.58 | 2.59 | 19.87 |
| 400mm girth | m | 1.10 | 15.40 | 3.44 | 2.83 | 21.67 |
| Aluminium capping to ridge | | | | | | |
| 0.6mm thick | | | | | | |
| 200mm girth | m | 0.85 | 11.90 | 1.50 | 2.01 | 15.41 |
| 300mm girth | m | 0.90 | 12.60 | 2.24 | 2.23 | 17.07 |
| 400mm girth | m | 0.95 | 13.30 | 2.99 | 2.44 | 18.73 |
| 0.9mm thick | | | | | | |
| 200mm girth | m | 0.85 | 11.90 | 1.72 | 2.04 | 15.66 |
| 300mm girth | m | 0.90 | 12.60 | 2.58 | 2.28 | 17.46 |
| 400mm girth | m | 0.95 | 13.30 | 3.44 | 2.51 | 19.25 |
| Aluminium capping to hip | | | | | | |
| 0.6mm thick | | | | | | |
| 200mm girth | m | 0.85 | 11.90 | 1.50 | 2.01 | 15.41 |
| 300mm girth | m | 0.90 | 12.60 | 2.24 | 2.23 | 17.07 |
| 400mm girth | m | 0.95 | 13.30 | 2.99 | 2.44 | 18.73 |
| 0.9mm thick | | | | | | |
| 200mm girth | m | 0.85 | 11.90 | 1.72 | 2.04 | 15.66 |
| 300mm girth | m | 0.90 | 12.60 | 2.58 | 2.28 | 17.46 |
| 400mm girth | m | 0.95 | 13.30 | 3.44 | 2.51 | 19.25 |

| | Unit | Labour | Hours £ | Mat'ls £ | O & P £ | Total £ |
|---|---|---|---|---|---|---|
| Aluminium lining to valley | | | | | | |
| 0.6mm thick | | | | | | |
| 200mm girth | m | 0.85 | 11.90 | 1.50 | 2.01 | 15.41 |
| 300mm girth | m | 0.90 | 12.60 | 2.24 | 2.23 | 17.07 |
| 400mm girth | m | 0.95 | 13.30 | 2.99 | 2.44 | 18.73 |
| 0.9mm thick | | | | | | |
| 200mm girth | m | 0.85 | 11.90 | 1.72 | 2.04 | 15.66 |
| 300mm girth | m | 0.90 | 12.60 | 2.58 | 2.28 | 17.46 |
| 400mm girth | m | 0.95 | 13.30 | 3.44 | 2.51 | 19.25 |
| Aluminium lining to gutter | | | | | | |
| 0.6mm thick | | | | | | |
| 200mm girth | m | 0.85 | 11.90 | 1.50 | 2.01 | 15.41 |
| 300mm girth | m | 0.90 | 12.60 | 2.24 | 2.23 | 17.07 |
| 400mm girth | m | 0.95 | 13.30 | 2.99 | 2.44 | 18.73 |
| 0.9mm thick | | | | | | |
| 200mm girth | m | 0.85 | 11.90 | 1.72 | 2.04 | 15.66 |
| 300mm girth | m | 0.90 | 12.60 | 2.58 | 2.28 | 17.46 |
| 400mm girth | m | 0.95 | 13.30 | 3.44 | 2.51 | 19.25 |
| Labours on aluminium | | | | | | |
| raking cutting | m | 0.10 | 1.40 | 0.00 | 0.21 | 1.61 |
| curved cutting | m | 0.15 | 2.10 | 0.00 | 0.32 | 2.42 |
| welted edge | m | 0.20 | 2.80 | 0.00 | 0.42 | 3.22 |
| beaded edge | m | 0.20 | 2.80 | 0.00 | 0.42 | 3.22 |
| wedging into groove | m | 0.20 | 2.80 | 0.00 | 0.42 | 3.22 |

| | Unit | Labour | Hours £ | Mat'ls £ | O & P £ | Total £ |
|---|---|---|---|---|---|---|
| **ZINC SHEET COVERINGS** | | | | | | |
| Zinc sheet coverings to roof with falls less than 10° | | | | | | |
| 0.65mm thick | m2 | 3.20 | 44.80 | 18.45 | 9.49 | 72.74 |
| 0.80mm thick | m2 | 3.20 | 44.80 | 20.37 | 9.78 | 74.95 |
| Zinc sheet coverings to roof with falls more than 10° and less that 50° | | | | | | |
| 0.65mm thick | m2 | 3.50 | 49.00 | 18.45 | 10.12 | 77.57 |
| 0.80mm thick | m2 | 3.50 | 49.00 | 20.37 | 10.41 | 79.78 |
| Zinc sheet coverings to roof with falls more than 50° | | | | | | |
| 0.65mm thick | m2 | 4.00 | 56.00 | 18.45 | 11.17 | 85.62 |
| 0.80mm thick | m2 | 4.00 | 56.00 | 20.37 | 11.46 | 87.83 |
| Zinc flashings | | | | | | |
| 0.65mm thick, horizontal | | | | | | |
| 150mm girth | m | 0.55 | 7.70 | 2.77 | 1.57 | 12.04 |
| 200mm girth | m | 0.65 | 9.10 | 3.69 | 1.92 | 14.71 |
| 300mm girth | m | 0.75 | 10.50 | 5.54 | 2.41 | 18.45 |
| 0.65mm thick, sloping | | | | | | |
| 150mm girth | m | 0.60 | 8.40 | 2.77 | 1.68 | 12.85 |
| 200mm girth | m | 0.70 | 9.80 | 3.69 | 2.02 | 15.51 |
| 300mm girth | m | 0.80 | 11.20 | 5.54 | 2.51 | 19.25 |
| 0.80mm thick, horizontal | | | | | | |
| 150mm girth | m | 0.55 | 7.70 | 3.06 | 1.61 | 12.37 |

| | Unit | Labour | Hours £ | Mat'ls £ | O & P £ | Total £ |
|---|---|---|---|---|---|---|
| **Zinc flashings (cont'd)** | | | | | | |
| 200mm girth | m | 0.65 | 9.10 | 4.07 | 1.98 | 15.15 |
| 300mm girth | m | 0.75 | 10.50 | 6.11 | 2.49 | 19.10 |
| 0.80mm thick, sloping | | | | | | |
| 150mm girth | m | 0.60 | 8.40 | 3.06 | 1.72 | 13.18 |
| 200mm girth | m | 0.70 | 9.80 | 4.07 | 2.08 | 15.95 |
| 300mm girth | m | 0.80 | 11.20 | 6.11 | 2.60 | 19.91 |
| Zinc stepped flashings | | | | | | |
| 0.65mm thick, sloping | | | | | | |
| 150mm girth | m | 0.70 | 9.80 | 2.77 | 1.89 | 14.46 |
| 200mm girth | m | 0.80 | 11.20 | 3.69 | 2.23 | 17.12 |
| 300mm girth | m | 0.90 | 12.60 | 5.54 | 2.72 | 20.86 |
| 0.80mm thick, sloping | | | | | | |
| 150mm girth | m | 0.70 | 9.80 | 3.06 | 1.93 | 14.79 |
| 200mm girth | m | 0.80 | 11.20 | 4.07 | 2.29 | 17.56 |
| 300mm girth | m | 0.90 | 12.60 | 6.11 | 2.81 | 21.52 |
| Zinc apron | | | | | | |
| 0.65mm thick | | | | | | |
| 200mm girth | m | 0.75 | 10.50 | 3.69 | 2.13 | 16.32 |
| 300mm girth | m | 0.80 | 11.20 | 5.54 | 2.51 | 19.25 |
| 400mm girth | m | 0.85 | 11.90 | 7.38 | 2.89 | 22.17 |
| 0.80mm thick | | | | | | |
| 200mm girth | m | 1.00 | 14.00 | 3.06 | 2.56 | 19.62 |
| 300mm girth | m | 1.05 | 14.70 | 4.07 | 2.82 | 21.59 |
| 400mm girth | m | 1.10 | 15.40 | 6.11 | 3.23 | 24.74 |

| | Unit | Labour | Hours £ | Mat'ls £ | O & P £ | Total £ |
|---|---|---|---|---|---|---|
| Zinc capping to ridge | | | | | | |
| 0.65mm thick | | | | | | |
| 200mm girth | m | 0.85 | 11.90 | 3.69 | 2.34 | 17.93 |
| 300mm girth | m | 0.90 | 12.60 | 5.54 | 2.72 | 20.86 |
| 400mm girth | m | 0.95 | 13.30 | 7.38 | 3.10 | 23.78 |
| 0.80mm thick | | | | | | |
| 200mm girth | m | 0.85 | 11.90 | 3.06 | 2.24 | 17.20 |
| 300mm girth | m | 0.90 | 12.60 | 4.07 | 2.50 | 19.17 |
| 400mm girth | m | 0.95 | 13.30 | 6.11 | 2.91 | 22.32 |
| Zinc capping to hip | | | | | | |
| 0.65mm thick | | | | | | |
| 200mm girth | m | 0.85 | 11.90 | 3.69 | 2.34 | 17.93 |
| 300mm girth | m | 0.90 | 12.60 | 5.54 | 2.72 | 20.86 |
| 400mm girth | m | 0.95 | 13.30 | 7.38 | 3.10 | 23.78 |
| 0.80mm thick | | | | | | |
| 200mm girth | m | 0.85 | 11.90 | 3.06 | 2.24 | 17.20 |
| 300mm girth | m | 0.90 | 12.60 | 4.07 | 2.50 | 19.17 |
| 400mm girth | m | 0.95 | 13.30 | 6.11 | 2.91 | 22.32 |
| Zinc lining to valley | | | | | | |
| 0.65mm thick | | | | | | |
| 200mm girth | m | 0.85 | 11.90 | 3.69 | 2.34 | 17.93 |
| 300mm girth | m | 0.90 | 12.60 | 5.54 | 2.72 | 20.86 |
| 400mm girth | m | 0.95 | 13.30 | 7.38 | 3.10 | 23.78 |
| 0.80mm thick | | | | | | |
| 200mm girth | m | 0.85 | 11.90 | 3.06 | 2.24 | 17.20 |
| 300mm girth | m | 0.90 | 12.60 | 4.07 | 2.50 | 19.17 |
| 400mm girth | m | 0.95 | 13.30 | 6.11 | 2.91 | 22.32 |

| | Unit | Labour | Hours £ | Mat'ls £ | O & P £ | Total £ |
|---|---|---|---|---|---|---|
| Zinc lining to gutter | | | | | | |
| 0.65mm thick | | | | | | |
| 200mm girth | m | 0.85 | 11.90 | 3.69 | 2.34 | 17.93 |
| 300mm girth | m | 0.90 | 12.60 | 5.54 | 2.72 | 20.86 |
| 400mm girth | m | 0.95 | 13.30 | 7.38 | 3.10 | 23.78 |
| 0.80mm thick | | | | | | |
| 200mm girth | m | 0.85 | 11.90 | 3.06 | 2.24 | 17.20 |
| 300mm girth | m | 0.90 | 12.60 | 4.07 | 2.50 | 19.17 |
| 400mm girth | m | 0.95 | 13.30 | 6.11 | 2.91 | 22.32 |
| Labours on zinc | | | | | | |
| raking cutting | m | 0.15 | 2.10 | 0.00 | 0.32 | 2.42 |
| curved cutting | m | 0.20 | 2.80 | 0.00 | 0.42 | 3.22 |
| welted edge | m | 0.25 | 3.50 | 0.00 | 0.53 | 4.03 |
| beaded edge | m | 0.25 | 3.50 | 0.00 | 0.53 | 4.03 |
| wedging into groove | m | 0.25 | 3.50 | 0.00 | 0.53 | 4.03 |

**BUILT-UP ROOFING**

| | Unit | Labour | Hours £ | Mat'ls £ | O & P £ | Total £ |
|---|---|---|---|---|---|---|
| Two layer built-up bituminous felt roof coverings, the layers bonded with hot bitumen | | | | | | |
| laid flat | m2 | 0.35 | 4.90 | 5.46 | 1.55 | 11.91 |
| laid sloping | m2 | 0.40 | 5.60 | 5.46 | 1.66 | 12.72 |
| Extra for | | | | | | |
| aprons | | | | | | |
| 100mm girth | m | 0.25 | 3.50 | 1.28 | 0.72 | 5.50 |
| 150mm girth | m | 0.30 | 4.20 | 1.84 | 0.91 | 6.95 |

| | Unit | Labour | Hours £ | Mat'ls £ | O & P £ | Total £ |
|---|---|---|---|---|---|---|
| skirtings | | | | | | |
| 100mm girth | m | 0.25 | 3.50 | 1.28 | 0.72 | 5.50 |
| 150mm girth | m | 0.30 | 4.20 | 1.84 | 0.91 | 6.95 |
| flashings | | | | | | |
| 100mm girth | m | 0.25 | 3.50 | 1.28 | 0.72 | 5.50 |
| 150mm girth | m | 0.30 | 4.20 | 1.84 | 0.91 | 6.95 |
| Form collars around 100mm diameter pipe | nr | 1.25 | 17.50 | 2.25 | 2.96 | 22.71 |
| Form collars around 150mm diameter pipe | nr | 1.35 | 18.90 | 4.08 | 3.45 | 26.43 |
| Three layer built-up bituminous felt roof coverings, the layers bonded with hot bitumen | | | | | | |
| laid flat | m2 | 0.50 | 7.00 | 8.46 | 2.32 | 17.78 |
| laid sloping | m2 | 0.55 | 7.70 | 8.46 | 2.42 | 18.58 |
| Extra for | | | | | | |
| aprons | | | | | | |
| 100mm girth | m | 0.30 | 4.20 | 1.51 | 0.86 | 6.57 |
| 150mm girth | m | 0.35 | 4.90 | 2.01 | 1.04 | 7.95 |
| skirtings | | | | | | |
| 100mm girth | m | 0.30 | 4.20 | 1.51 | 0.86 | 6.57 |
| 150mm girth | m | 0.35 | 4.90 | 2.01 | 1.04 | 7.95 |
| flashings | | | | | | |
| 100mm girth | m | 0.30 | 4.20 | 1.51 | 0.86 | 6.57 |
| 150mm girth | m | 0.35 | 4.90 | 2.01 | 1.04 | 7.95 |

| | Unit | Labour | Hours £ | Mat'ls £ | O & P £ | Total £ |
|---|---|---|---|---|---|---|
| Form collars around 100mm diameter pipe | nr | 1.50 | 21.00 | 3.62 | 3.69 | 28.31 |
| Form collars around 150mm diameter pipe | nr | 1.60 | 22.40 | 4.08 | 3.97 | 30.45 |

**RAINWATER PIPES**

PVC-U rainwater pipe plugged to brickwork with pipe brackets and fitting clips at 2m maximum centres

| | Unit | Labour | Hours £ | Mat'ls £ | O & P £ | Total £ |
|---|---|---|---|---|---|---|
| 68mm diameter pipe | m | 0.25 | 3.50 | 5.83 | 1.40 | 10.73 |
| extra over for | | | | | | |
| bend, 87.5° | nr | 0.25 | 3.50 | 5.60 | 1.37 | 10.47 |
| offset | nr | 0.25 | 3.50 | 3.33 | 1.02 | 7.85 |
| branch | nr | 0.25 | 3.50 | 11.15 | 2.20 | 16.85 |
| shoe | nr | 0.25 | 3.50 | 4.83 | 1.25 | 9.58 |
| access pipe | nr | 0.25 | 3.50 | 15.36 | 2.83 | 21.69 |
| hopper head | nr | 0.25 | 3.50 | 17.95 | 3.22 | 24.67 |
| 68mm square pipe | m | 0.25 | 3.50 | 6.98 | 1.57 | 12.05 |
| extra over for | | | | | | |
| bend, 87.5° | nr | 0.25 | 3.50 | 3.57 | 1.06 | 8.13 |
| offset | nr | 0.25 | 3.50 | 6.28 | 1.47 | 11.25 |
| branch | nr | 0.25 | 3.50 | 12.64 | 2.42 | 18.56 |
| shoe | nr | 0.25 | 3.50 | 4.10 | 1.14 | 8.74 |
| access pipe | nr | 0.25 | 3.50 | 19.17 | 3.40 | 26.07 |
| hopper head | nr | 0.25 | 3.50 | 17.95 | 3.22 | 24.67 |

| | Unit | Labour | Hours £ | Mat'ls £ | O & P £ | Total £ |
|---|---|---|---|---|---|---|
| Aluminium rainwater pipe, straight, plain eared, spigot and socket dry joints plugged to brickwork | | | | | | |
| 63mm diameter pipe | m | 0.28 | 3.92 | 15.94 | 2.98 | 22.84 |
| extra over for | | | | | | |
| offset, 75mm | nr | 0.28 | 3.92 | 30.17 | 5.11 | 39.20 |
| offset, 100mm | nr | 0.28 | 3.92 | 30.17 | 5.11 | 39.20 |
| offset, 150mm | nr | 0.28 | 3.92 | 30.17 | 5.11 | 39.20 |
| offset, 225mm | nr | 0.28 | 3.92 | 36.21 | 6.02 | 46.15 |
| offset, 300mm | nr | 0.28 | 3.92 | 36.21 | 6.02 | 46.15 |
| bend, 92.5° | nr | 0.28 | 3.92 | 24.55 | 4.27 | 32.74 |
| branch | nr | 0.28 | 3.92 | 26.27 | 4.53 | 34.72 |
| shoe | nr | 0.28 | 3.92 | 9.15 | 1.96 | 15.03 |
| standard hopper | nr | 0.28 | 3.92 | 25.51 | 4.41 | 33.84 |
| flatback hopper | nr | 0.28 | 3.92 | 17.82 | 3.26 | 25.00 |
| 76mm diameter pipe | m | 0.30 | 4.20 | 19.64 | 3.58 | 27.42 |
| extra over for | | | | | | |
| offset, 75mm | nr | 0.30 | 4.20 | 37.11 | 6.20 | 47.51 |
| offset, 100mm | nr | 0.30 | 4.20 | 37.11 | 6.20 | 47.51 |
| offset, 150mm | nr | 0.30 | 4.20 | 37.11 | 6.20 | 47.51 |
| offset, 225mm | nr | 0.30 | 4.20 | 42.96 | 7.07 | 54.23 |
| offset, 300mm | nr | 0.30 | 4.20 | 44.53 | 7.31 | 56.04 |
| bend, 92.5° | nr | 0.30 | 4.20 | 27.60 | 4.77 | 36.57 |
| branch | nr | 0.30 | 4.20 | 31.56 | 5.36 | 41.12 |
| shoe | nr | 0.30 | 4.20 | 12.64 | 2.53 | 19.37 |
| standard hopper | nr | 0.30 | 4.20 | 26.68 | 4.63 | 35.51 |
| flatback hopper | nr | 0.30 | 4.20 | 17.82 | 3.30 | 25.32 |

| | Unit | Labour | Hours £ | Mat'ls £ | O & P £ | Total £ |
|---|---|---|---|---|---|---|
| 102mm diameter pipe | m | 0.32 | 4.48 | 27.43 | 4.79 | 36.70 |
| extra over for | | | | | | |
| offset, 75mm | nr | 0.32 | 4.48 | 45.12 | 7.44 | 57.04 |
| offset, 100mm | nr | 0.32 | 4.48 | 45.12 | 7.44 | 57.04 |
| offset, 150mm | nr | 0.32 | 4.48 | 45.12 | 7.44 | 57.04 |
| offset, 225mm | nr | 0.32 | 4.48 | 52.65 | 8.57 | 65.70 |
| offset, 300mm | nr | 0.32 | 4.48 | 52.65 | 8.57 | 65.70 |
| bend, 92.5° | nr | 0.32 | 4.48 | 39.57 | 6.61 | 50.66 |
| branch | nr | 0.32 | 4.48 | 42.44 | 7.04 | 53.96 |
| shoe | nr | 0.32 | 4.48 | 16.69 | 3.18 | 24.35 |
| standard hopper | nr | 0.32 | 4.48 | 35.72 | 6.03 | 46.23 |
| ornamental hopper | nr | 0.32 | 4.48 | 124.80 | 19.39 | 148.67 |
| Cast iron rainwater pipe, straight, ears cast on, spigot and socket dry joints plugged to brickwork | | | | | | |
| 65mm diameter pipe | m | 0.25 | 3.50 | 21.28 | 3.72 | 28.50 |
| extra over for | | | | | | |
| offset, 75mm | nr | 0.25 | 3.50 | 15.89 | 2.91 | 22.30 |
| offset, 115mm | nr | 0.25 | 3.50 | 15.89 | 2.91 | 22.30 |
| offset, 150mm | nr | 0.25 | 3.50 | 15.89 | 2.91 | 22.30 |
| offset, 225mm | nr | 0.25 | 3.50 | 18.50 | 3.30 | 25.30 |
| offset, 305mm | nr | 0.25 | 3.50 | 21.66 | 3.77 | 28.93 |
| bend | nr | 0.25 | 3.50 | 10.37 | 2.08 | 15.95 |
| branch | nr | 0.25 | 3.50 | 20.40 | 3.59 | 27.49 |
| shoe | nr | 0.25 | 3.50 | 14.70 | 2.73 | 20.93 |
| eared shoe | nr | 0.25 | 3.50 | 16.95 | 3.07 | 23.52 |
| flat hopper | nr | 0.25 | 3.50 | 13.24 | 2.51 | 19.25 |
| rectangular hopper | nr | 0.25 | 3.50 | 59.00 | 9.38 | 71.88 |

| | Unit | Labour | Hours £ | Mat'ls £ | O & P £ | Total £ |
|---|---|---|---|---|---|---|
| 75mm diameter pipe | m | 0.30 | 4.20 | 21.33 | 3.83 | 29.36 |
| extra over for | | | | | | |
| offset, 75mm | nr | 0.30 | 4.20 | 15.89 | 3.01 | 23.10 |
| offset, 115mm | nr | 0.30 | 4.20 | 15.89 | 3.01 | 23.10 |
| offset, 150mm | nr | 0.30 | 4.20 | 15.89 | 3.01 | 23.10 |
| offset, 225mm | nr | 0.30 | 4.20 | 18.50 | 3.41 | 26.11 |
| offset, 305mm | nr | 0.30 | 4.20 | 22.73 | 4.04 | 30.97 |
| bend | nr | 0.30 | 4.20 | 12.60 | 2.52 | 19.32 |
| branch | nr | 0.30 | 4.20 | 22.06 | 3.94 | 30.20 |
| shoe | nr | 0.30 | 4.20 | 14.70 | 2.84 | 21.74 |
| eared shoe | nr | 0.30 | 4.20 | 16.95 | 3.17 | 24.32 |
| flat hopper | nr | 0.30 | 4.20 | 15.04 | 2.89 | 22.13 |
| rectangular hopper | nr | 0.30 | 4.20 | 59.00 | 9.48 | 72.68 |
| 100mm diameter pipe | m | 0.35 | 4.90 | 27.87 | 4.92 | 37.69 |
| extra over for | | | | | | |
| offset, 75mm | nr | 0.35 | 4.90 | 29.98 | 5.23 | 40.11 |
| offset, 115mm | nr | 0.35 | 4.90 | 29.98 | 5.23 | 40.11 |
| offset, 150mm | nr | 0.35 | 4.90 | 29.98 | 5.23 | 40.11 |
| offset, 225mm | nr | 0.35 | 4.90 | 36.30 | 6.18 | 47.38 |
| offset, 305mm | nr | 0.35 | 4.90 | 59.00 | 9.59 | 73.49 |
| bend | nr | 0.35 | 4.90 | 17.80 | 3.41 | 26.11 |
| branch | nr | 0.35 | 4.90 | 26.21 | 4.67 | 35.78 |
| shoe | nr | 0.35 | 4.90 | 19.82 | 3.71 | 28.43 |
| eared shoe | nr | 0.35 | 4.90 | 22.52 | 4.11 | 31.53 |
| flat hopper | nr | 0.35 | 4.90 | 33.34 | 5.74 | 43.98 |
| rectangular hopper | nr | 0.35 | 4.90 | 59.00 | 9.59 | 73.49 |

## RAINWATER GUTTERS

PVC-U half round rainwater gutter, fixed to timber with support brackets at 1m maximum centres

| | Unit | Labour | Hours<br>£ | Mat'ls<br>£ | O & P<br>£ | Total<br>£ |
|---|---|---|---|---|---|---|
| 76mm elliptical gutter | m | 0.20 | 2.80 | 6.31 | 1.37 | 10.48 |
| extra over for | | | | | | |
| running outlet | nr | 0.20 | 2.80 | 5.75 | 1.28 | 9.83 |
| angle | nr | 0.20 | 2.80 | 6.27 | 1.36 | 10.43 |
| stop end outlet | nr | 0.10 | 1.40 | 5.04 | 0.97 | 7.41 |
| stop end | nr | 0.10 | 1.40 | 3.08 | 0.67 | 5.15 |
| 112mm half-round gutter | m | 0.26 | 3.64 | 4.65 | 1.24 | 9.53 |
| extra over for | | | | | | |
| running outlet | nr | 0.26 | 3.64 | 4.05 | 1.15 | 8.84 |
| angle | nr | 0.26 | 3.64 | 4.62 | 1.24 | 9.50 |
| stop end outlet | nr | 0.13 | 1.82 | 4.05 | 0.88 | 6.75 |
| stop end | nr | 0.13 | 1.82 | 2.02 | 0.58 | 4.42 |
| 150mm half-round gutter | m | 0.30 | 4.20 | 6.87 | 1.66 | 12.73 |
| extra over for | | | | | | |
| running outlet | nr | 0.30 | 4.20 | 9.36 | 2.03 | 15.59 |
| angle | nr | 0.30 | 4.20 | 9.64 | 2.08 | 15.92 |
| stop end outlet | nr | 0.15 | 2.10 | 5.88 | 1.20 | 9.18 |
| stop end | nr | 0.15 | 2.10 | 1.89 | 0.60 | 4.59 |
| Aluminium half round rainwater gutter with mastic joints, fixed to timber with support brackets at 1m maximum centres | | | | | | |
| 100mm gutter | m | 0.30 | 4.20 | 12.78 | 2.55 | 19.53 |
| extra over for | | | | | | |
| running outlet | nr | 0.30 | 4.20 | 11.39 | 2.34 | 17.93 |
| angle | nr | 0.30 | 4.20 | 6.70 | 1.64 | 12.54 |
| stop end outlet | nr | 0.15 | 2.10 | 6.55 | 1.30 | 9.95 |
| stop end | nr | 0.15 | 2.10 | 2.52 | 0.69 | 5.31 |

| | Unit | Labour | Hours £ | Mat'ls £ | O & P £ | Total £ |
|---|---|---|---|---|---|---|
| 125mm gutter | m | 0.32 | 4.48 | 16.10 | 3.09 | 23.67 |
| extra over for | | | | | | |
| running outlet | nr | 0.32 | 4.48 | 10.89 | 2.31 | 17.68 |
| angle | nr | 0.32 | 4.48 | 9.14 | 2.04 | 15.66 |
| stop end outlet | nr | 0.16 | 2.24 | 7.54 | 1.47 | 11.25 |
| stop end | nr | 0.16 | 2.24 | 3.28 | 0.83 | 6.35 |
| Aluminium ogee rainwater gutter with mastic joints, fixed to timber with support brackets at 1m maximum centres | | | | | | |
| 120 × 75mm gutter | m | 0.30 | 4.20 | 16.51 | 3.11 | 23.82 |
| extra over for | | | | | | |
| running outlet | nr | 0.30 | 4.20 | 20.80 | 3.75 | 28.75 |
| angle | nr | 0.30 | 4.20 | 17.42 | 3.24 | 24.86 |
| stop end outlet | nr | 0.15 | 2.10 | 29.06 | 4.67 | 35.83 |
| stop end | nr | 0.15 | 2.10 | 4.05 | 0.92 | 7.07 |
| 155 × 100mm gutter | m | 0.32 | 4.48 | 31.14 | 5.34 | 40.96 |
| extra over for | | | | | | |
| running outlet | nr | 0.32 | 4.48 | 27.50 | 4.80 | 36.78 |
| angle | nr | 0.32 | 4.48 | 26.89 | 4.71 | 36.08 |
| stop end outlet | nr | 0.16 | 2.24 | 22.41 | 3.70 | 28.35 |
| stop end | nr | 0.16 | 2.24 | 16.36 | 2.79 | 21.39 |
| Cast iron half-round rainwater gutter with mastic joints, primed, fixed to timber with support brackets at 1m maximum centres | | | | | | |
| 100mm gutter | m | 0.36 | 5.04 | 11.28 | 2.45 | 18.77 |
| extra over for | | | | | | |
| running outlet | nr | 0.36 | 5.04 | 7.24 | 1.84 | 14.12 |

| | Unit | Labour | Hours £ | Mat'ls £ | O & P £ | Total £ |
|---|---|---|---|---|---|---|
| **Cast iron half-round rainwater gutter (cont'd)** | | | | | | |
| angle | nr | 0.36 | 5.04 | 7.44 | 1.87 | 14.35 |
| stop end outlet | nr | 0.18 | 2.52 | 8.50 | 1.65 | 12.67 |
| stop end | nr | 0.18 | 2.52 | 2.50 | 0.75 | 5.77 |
| 115mm gutter | m | 0.38 | 5.32 | 11.64 | 2.54 | 19.50 |
| extra over for | | | | | | |
| running outlet | nr | 0.38 | 5.32 | 7.90 | 1.98 | 15.20 |
| angle | nr | 0.38 | 5.32 | 7.66 | 1.95 | 14.93 |
| stop end outlet | nr | 0.19 | 2.66 | 9.53 | 1.83 | 14.02 |
| stop end | nr | 0.19 | 2.66 | 3.24 | 0.89 | 6.79 |
| 125mm gutter | m | 0.40 | 5.60 | 13.46 | 2.86 | 21.92 |
| extra over for | | | | | | |
| running outlet | nr | 0.40 | 5.60 | 7.51 | 1.97 | 15.08 |
| angle | nr | 0.40 | 5.60 | 9.03 | 2.19 | 16.82 |
| stop end outlet | nr | 0.20 | 2.80 | 10.87 | 2.05 | 15.72 |
| stop end | nr | 0.20 | 2.80 | 3.24 | 0.91 | 6.95 |
| 150mm gutter | m | 0.44 | 6.16 | 22.50 | 4.30 | 32.96 |
| extra over for | | | | | | |
| running outlet | nr | 0.44 | 6.16 | 14.10 | 3.04 | 23.30 |
| angle | nr | 0.44 | 6.16 | 16.50 | 3.40 | 26.06 |
| stop end outlet | nr | 0.22 | 3.08 | 17.88 | 3.14 | 24.10 |
| stop end | nr | 0.22 | 3.08 | 4.99 | 1.21 | 9.28 |
| **ROOF OUTLETS** | | | | | | |
| Cast iron circular roof outlet with flat grate, diameter | | | | | | |
| 50mm | nr | 0.50 | 7.00 | 77.59 | 12.69 | 97.28 |
| 75mm | nr | 0.60 | 8.40 | 84.59 | 13.95 | 106.94 |
| 100mm | nr | 0.70 | 9.80 | 99.15 | 16.34 | 125.29 |

| | Unit | Labour | Hours £ | Mat'ls £ | O & P £ | Total £ |
|---|---|---|---|---|---|---|
| Aluminium circular roof outlet with domed grate, diameter | | | | | | |
| 50mm | nr | 0.50 | 7.00 | 71.60 | 11.79 | 90.39 |
| 75mm | nr | 0.60 | 8.40 | 81.74 | 13.52 | 103.66 |
| 100mm | nr | 0.70 | 9.80 | 105.51 | 17.30 | 132.61 |
| 150mm | nr | 0.80 | 11.20 | 118.09 | 19.39 | 148.68 |
| Plastic wire balloon guard for pipes and outlets, diameter | | | | | | |
| 50mm | nr | 0.05 | 0.70 | 2.05 | 0.41 | 3.16 |
| 63mm | nr | 0.05 | 0.70 | 2.18 | 0.43 | 3.31 |
| 75mm | nr | 0.05 | 0.70 | 2.24 | 0.44 | 3.38 |
| 100mm | nr | 0.05 | 0.70 | 2.26 | 0.44 | 3.40 |
| Galvanised wire balloon guard for pipes and outlets, diameter | | | | | | |
| 50mm | nr | 0.05 | 0.70 | 1.71 | 0.36 | 2.77 |
| 63mm | nr | 0.05 | 0.70 | 1.77 | 0.37 | 2.84 |
| 75mm | nr | 0.05 | 0.70 | 1.82 | 0.38 | 2.90 |
| 100mm | nr | 0.05 | 0.70 | 2.26 | 0.44 | 3.40 |
| Copper wire balloon guard for pipes and outlets, diameter | | | | | | |
| 50mm | nr | 0.05 | 0.70 | 2.04 | 0.41 | 3.15 |
| 63mm | nr | 0.05 | 0.70 | 2.08 | 0.42 | 3.20 |
| 75mm | nr | 0.05 | 0.70 | 2.23 | 0.44 | 3.37 |
| 100mm | nr | 0.05 | 0.70 | 3.14 | 0.58 | 4.42 |

| | Unit | Labour | Hours £ | Mat'ls £ | O & P £ | Total £ |
|---|---|---|---|---|---|---|
| **ROOF LIGHTS** | | | | | | |
| Roof windows, Velux centre-pivot GGL range, double glazing, clear lacquer finish, reference | | | | | | |
| GGL 3000 CO2, size 550 x 780mm | nr | 2.30 | 32.20 | 126.00 | 23.73 | 181.93 |
| GGL 3000 CO4, size 550 x 980mm | nr | 2.30 | 32.20 | 140.00 | 25.83 | 198.03 |
| GGL 3000 FO6, size 660 x 1180mm | nr | 2.50 | 35.00 | 170.00 | 30.75 | 235.75 |
| GGL 3000 MO4, size 780 x 980mm | nr | 2.50 | 35.00 | 160.00 | 29.25 | 224.25 |
| GGL 3000 MO6, size 780 x 1180mm | nr | 2.50 | 35.00 | 182.00 | 32.55 | 249.55 |
| GGL 3000 MO8, size 780 x 1400mm | nr | 3.00 | 42.00 | 198.00 | 36.00 | 276.00 |
| GGL 3000 P10, size 940 x 1600mm | nr | 3.30 | 46.20 | 240.00 | 42.93 | 329.13 |
| GGL 3000 SO6, size 1140 x 1180mm | nr | 3.00 | 42.00 | 220.00 | 39.30 | 301.30 |
| Roof windows, Velux centre-pivot GGL range, double glazing with toughened glass, clear lacquer finish, reference | | | | | | |
| GGL 3073 CO2, size 550 x 780mm | nr | 2.30 | 32.20 | 165.00 | 29.58 | 226.78 |
| GGL 3073 CO4, size 550 x 980mm | nr | 2.30 | 32.20 | 182.00 | 32.13 | 246.33 |
| GGL 3073 FO6, size 660 x 1180mm | nr | 2.50 | 35.00 | 218.00 | 37.95 | 290.95 |

| | Unit | Labour | Hours £ | Mat'ls £ | O & P £ | Total £ |
|---|---|---|---|---|---|---|
| GGL 3073 MO4, size 780 x 980mm | nr | 2.50 | 35.00 | 208.00 | 36.45 | 279.45 |
| GGL 3073 MO6, size 780 x 1180mm | nr | 2.50 | 35.00 | 231.00 | 39.90 | 305.90 |
| GGL 3073 MO8, size 780 x 1400mm | nr | 3.00 | 42.00 | 255.00 | 44.55 | 341.55 |
| GGL 3073 P10, size 940 x 1600mm | nr | 3.30 | 46.20 | 312.00 | 53.73 | 411.93 |
| GGL 3073 SO6, size 1140 x 1180mm | nr | 3.00 | 42.00 | 286.00 | 49.20 | 377.20 |
| Roof windows, Velux centre-pivot GGU range, insulated double glazing, white poly-urethane finish, reference | | | | | | |
| GGU 0059 CO2, size 550 x 780mm | nr | 2.30 | 32.20 | 170.00 | 30.33 | 232.53 |
| GGU 0059 CO4, size 550 x 980mm | nr | 2.30 | 32.20 | 180.00 | 31.83 | 244.03 |
| GGU 0059 FO6, size 660 x 1180mm | nr | 2.50 | 35.00 | 212.00 | 37.05 | 284.05 |
| GGU 0059 MO4, size 780 x 980mm | nr | 2.50 | 35.00 | 204.00 | 35.85 | 274.85 |
| GGU 0059 MO6, size 780 x 1180mm | nr | 2.50 | 35.00 | 222.00 | 38.55 | 295.55 |
| GGU 0059 MO8, size 780 x 1400mm | nr | 3.00 | 42.00 | 248.00 | 43.50 | 333.50 |
| GGU 0059 SO6, size 1140 x 1180mm | nr | 3.00 | 42.00 | 275.00 | 47.55 | 364.55 |

| | Unit | Labour | Hours £ | Mat'ls £ | O & P £ | Total £ |
|---|---|---|---|---|---|---|
| Roof windows, Velux centre-pivot GGU range, insulated double glazing with toughened glass, inner skin obscure glass, white polyurethane finish, reference | | | | | | |
| GGU 0034 CO2, size 550 x 780mm | nr | 2.30 | 32.20 | 178.00 | 31.53 | 241.73 |
| GGU 0059 CO4, size 550 x 980mm | nr | 2.30 | 32.20 | 180.00 | 31.83 | 244.03 |
| GGU 0034 FO6, size 660 x 1180mm | nr | 2.50 | 35.00 | 210.00 | 36.75 | 281.75 |
| GGU 0034 MO4, size 780 x 980mm | nr | 2.50 | 35.00 | 205.00 | 36.00 | 276.00 |
| GGU 0034 MO6, size 780 x 1180mm | nr | 2.50 | 35.00 | 222.00 | 38.55 | 295.55 |
| GGU 0034 MO8, size 780 x 1400mm | nr | 3.00 | 42.00 | 247.00 | 43.35 | 332.35 |
| GGU 0034 SO6, size 1140 x 1180mm | nr | 3.00 | 42.00 | 273.00 | 47.25 | 362.25 |
| Roof windows, Velux centre-pivot GGU range, insulated double glazing, toughened glass, white polyurethane finish, reference | | | | | | |
| GGU 0073 CO2, size 550 x 780mm | nr | 2.30 | 32.20 | 210.00 | 36.33 | 278.53 |
| GGU 0073 CO4, size 550 x 980mm | nr | 2.30 | 32.20 | 222.00 | 38.13 | 292.33 |
| GGU 0073 FO6, size 660 x 1180mm | nr | 2.50 | 35.00 | 266.00 | 45.15 | 346.15 |
| GGU 0073 MO4, size 780 x 980mm | nr | 2.50 | 35.00 | 258.00 | 43.95 | 336.95 |

| | Unit | Labour | Hours £ | Mat'ls £ | O & P £ | Total £ |
|---|---|---|---|---|---|---|
| GGU 0073 MO6, size 780 x 1180mm | nr | 2.50 | 35.00 | 282.00 | 47.55 | 364.55 |
| GGU 0073 MO8, size 780 x 1400mm | nr | 3.00 | 42.00 | 318.00 | 54.00 | 414.00 |
| GGU 0073 SO6, size 1140 x 1180mm | nr | 3.00 | 42.00 | 354.00 | 59.40 | 455.40 |
| Roof windows, Velux top hung GHL range, insulated double glazing, clear lacquer finish, reference | | | | | | |
| GHL 3059 CO4, size 550 x 980mm | nr | 2.30 | 32.20 | 196.00 | 34.23 | 262.43 |
| GHL 3059 FO6, size 660 x 1180mm | nr | 2.50 | 35.00 | 225.00 | 39.00 | 299.00 |
| GGU 0073 MO4, size 780 x 980mm | nr | 2.50 | 35.00 | 216.00 | 37.65 | 288.65 |
| GHL 3059 MO6, size 780 x 1180mm | nr | 2.50 | 35.00 | 238.00 | 40.95 | 313.95 |
| GHL 3059 MO8, size 780 x 1400mm | nr | 3.00 | 42.00 | 258.00 | 45.00 | 345.00 |
| GHL 3059 SO6, size 1140 x 1180mm | nr | 3.00 | 42.00 | 274.00 | 47.40 | 363.40 |
| Roof windows, Velux top hung GHL range, double glazing with toughened glass, clear lacquer finish, reference | | | | | | |
| GHL 3073 CO4, size 550 x 980mm | nr | 2.30 | 32.20 | 235.00 | 40.08 | 307.28 |
| GHL 3073 FO6, size 660 x 1180mm | nr | 2.50 | 35.00 | 259.00 | 44.10 | 338.10 |

| | Unit | Labour | Hours £ | Mat'ls £ | O & P £ | Total £ |
|---|---|---|---|---|---|---|
| **Roof windows GHL range (cont'd)** | | | | | | |
| GGU 3073 MO4, size 780 x 980mm | nr | 2.50 | 35.00 | 249.00 | 42.60 | 326.60 |
| GHL 3073 MO6, size 780 x 1180mm | nr | 2.50 | 35.00 | 284.00 | 47.85 | 366.85 |
| GHL 3073 MO8, size 780 x 1400mm | nr | 3.00 | 42.00 | 320.00 | 54.30 | 416.30 |
| GHL 3073 SO6, size 1140 x 1180mm | nr | 3.00 | 42.00 | 345.00 | 58.05 | 445.05 |
| Roof windows, Velux top hung GPU range, insulated double glazing, white polyurethane finish, reference | | | | | | |
| GPU 0059 FO6, size 660 x 1180mm | nr | 2.50 | 35.00 | 284.00 | 47.85 | 366.85 |
| GPU 0059 MO6, size 780 x 1180mm | nr | 2.50 | 35.00 | 300.00 | 50.25 | 385.25 |
| GPU 0059 MO8, size 780 x 1400mm | nr | 3.00 | 42.00 | 322.00 | 54.60 | 418.60 |
| GPU 0059 SO6, size 1140 x 1180mm | nr | 3.00 | 42.00 | 345.00 | 58.05 | 445.05 |
| Roof windows, Velux top hung GPU range, toughened double glazing, inner leaf obscure glass, white polyurethane finish, reference | | | | | | |
| GPU 0034 FO6, size 660 x 1180mm | nr | 2.50 | 35.00 | 278.00 | 46.95 | 359.95 |
| GPU 0034 MO6, size 780 x 1180mm | nr | 2.50 | 35.00 | 288.00 | 48.45 | 371.45 |

| | Unit | Labour | Hours £ | Mat'ls £ | O & P £ | Total £ |
|---|---|---|---|---|---|---|
| GPU 0034 MO8, size 780 x 1400mm | nr | 3.00 | 42.00 | 316.00 | 53.70 | 411.70 |
| GPU 0034 SO6, size 1140 x 1180mm | nr | 3.00 | 42.00 | 338.00 | 57.00 | 437.00 |
| Roof windows, Velux centre pivot GPU range, toughened double glazing, white poly-urethane finish, reference | | | | | | |
| GPU 0073 FO6, size 660 x 980mm | nr | 2.50 | 35.00 | 352.00 | 58.05 | 445.05 |
| GPU 0073 MO6, size 780 x 1180mm | nr | 2.50 | 35.00 | 365.00 | 60.00 | 460.00 |
| GPU 0073 MO8, size 780 x 1400mm | nr | 3.00 | 42.00 | 400.00 | 66.30 | 508.30 |
| GPU 0073 SO6, size 1140 x 1180mm | nr | 3.00 | 42.00 | 452.00 | 74.10 | 568.10 |
| Roof windows, Velux centre pivot ACS conservation range, toughened double glazing, clear lacquer finish, reference | | | | | | |
| ACS 3573N CO4, size 550 x 980mm | nr | 2.50 | 35.00 | 365.00 | 60.00 | 460.00 |
| ACS 3573N FO6, size 660 x 1180mm | nr | 2.50 | 35.00 | 412.00 | 67.05 | 514.05 |
| ACS 3573N MO8, size 780 x 1400mm | nr | 3.00 | 42.00 | 520.00 | 84.30 | 646.30 |
| ACS 3573P CO4, size 550 x 980mm | nr | 2.50 | 35.00 | 365.00 | 60.00 | 460.00 |
| ACS 3573P FO6, size 660 x 1180mm | nr | 2.50 | 35.00 | 412.00 | 67.05 | 514.05 |
| ACS 3573P MO8, size 780 x 1400mm | nr | 3.00 | 42.00 | 520.00 | 84.30 | 646.30 |

| | Unit | Labour | Hours £ | Mat'ls £ | O & P £ | Total £ |
|---|---|---|---|---|---|---|
| **Roof windows ACS range (cont'd)** | | | | | | |
| ACS 3573H CO4, size 550 x 980mm | nr | 2.50 | 35.00 | 365.00 | 60.00 | 460.00 |
| ACS 3573H FO6, size 660 x 1180mm | nr | 2.50 | 35.00 | 412.00 | 67.05 | 514.05 |
| ACS 3573H MO8, size 780 x 1400mm | nr | 3.00 | 42.00 | 520.00 | 84.30 | 646.30 |
| Roof windows, Velux top hung ACE conservation range, toughened double glazing, clear lacquer finish, reference | | | | | | |
| ACE 3573N MO8, size 780 x 1400mm | nr | 3.00 | 42.00 | 642.00 | 102.60 | 786.60 |
| ACE 3573P MO8, size 780 x 1400mm | nr | 3.00 | 42.00 | 642.00 | 102.60 | 786.60 |
| ACE 3573H MO8, size 780 x 1400mm | nr | 3.00 | 42.00 | 642.00 | 102.60 | 786.60 |
| Aluminium flashing EDL 0000 for window | | | | | | |
| reference CO2, size 550 x 780mm | nr | 0.45 | 6.30 | 28.50 | 5.22 | 40.02 |
| reference CO4, size 550 x 080mm | nr | 0.45 | 6.30 | 30.25 | 5.48 | 42.03 |
| reference FO6, size 660 x 1180mm | nr | 1.00 | 14.00 | 34.56 | 7.28 | 55.84 |
| reference MO4, size 780 x 980mm | nr | 1.00 | 14.00 | 33.42 | 7.11 | 54.53 |
| reference MO6, size 780 x 1180mm | nr | 1.00 | 14.00 | 35.14 | 7.37 | 56.51 |
| reference MO8, size 780 x 140mm | nr | 1.00 | 14.00 | 36.47 | 7.57 | 58.04 |

| | Unit | Labour | Hours £ | Mat'ls £ | O & P £ | Total £ |
|---|---|---|---|---|---|---|
| reference P10, size 940 x 1600mm | nr | 1.10 | 15.40 | 40.22 | 8.34 | 63.96 |
| reference SO6, size 1140 x 1180mm | nr | 1.10 | 15.40 | 38.96 | 8.15 | 62.51 |
| Aluminium flashing EDN 0000 for window | | | | | | |
| reference CO2, size 550 x 780mm | nr | 0.45 | 6.30 | 45.24 | 7.73 | 59.27 |
| reference CO4, size 550 x 080mm | nr | 0.45 | 6.30 | 57.10 | 9.51 | 72.91 |
| reference FO6, size 660 x 1180mm | nr | 1.00 | 14.00 | 61.48 | 11.32 | 86.80 |
| reference MO4, size 780 x 980mm | nr | 1.00 | 14.00 | 59.60 | 11.04 | 84.64 |
| reference MO6, size 780 x 1180mm | nr | 1.00 | 14.00 | 65.14 | 11.87 | 91.01 |
| reference MO8, size 780 x 140mm | nr | 1.00 | 14.00 | 68.97 | 12.45 | 95.42 |
| reference P10, size 940 x 1600mm | nr | 1.10 | 15.40 | 72.57 | 13.20 | 101.17 |
| reference SO6, size 1140 x 1180mm | nr | 1.10 | 15.40 | 68.75 | 12.62 | 96.77 |
| Aluminium flashing EDZ 0000 for window | | | | | | |
| reference CO2, size 550 x 780mm | nr | 0.45 | 6.30 | 32.44 | 5.81 | 44.55 |
| reference CO4, size 550 x 080mm | nr | 0.45 | 6.30 | 34.82 | 6.17 | 47.29 |
| reference FO6, size 660 x 1180mm | nr | 1.00 | 14.00 | 37.64 | 7.75 | 59.39 |
| reference MO4, size 780 x 980mm | nr | 1.00 | 14.00 | 36.49 | 7.57 | 58.06 |

| | Unit | Labour | Hours £ | Mat'ls £ | O & P £ | Total £ |
|---|---|---|---|---|---|---|
| **Flashing EDZ range (cont'd)** | | | | | | |
| reference MO6, size 780 x 1180mm | nr | 1.00 | 14.00 | 39.17 | 7.98 | 61.15 |
| reference MO8, size 780 x 140mm | nr | 1.00 | 14.00 | 41.62 | 8.34 | 63.96 |
| reference P10, size 940 x 1600mm | nr | 1.10 | 15.40 | 44.67 | 9.01 | 69.08 |
| reference SO6, size 1140 x 1180mm | nr | 1.10 | 15.40 | 43.17 | 8.79 | 67.36 |
| Aluminium flashing EDH 0000 for window | | | | | | |
| reference CO2, size 550 x 780mm | nr | 0.45 | 6.30 | 37.88 | 6.63 | 50.81 |
| reference CO4, size 550 x 080mm | nr | 0.45 | 6.30 | 38.96 | 6.79 | 52.05 |
| reference FO6, size 660 x 1180mm | nr | 1.00 | 14.00 | 43.61 | 8.64 | 66.25 |
| reference MO4, size 780 x 980mm | nr | 1.00 | 14.00 | 43.61 | 8.64 | 66.25 |
| reference MO6, size 780 x 1180mm | nr | 1.00 | 14.00 | 45.72 | 8.96 | 68.68 |
| reference MO8, size 780 x 140mm | nr | 1.00 | 14.00 | 45.72 | 8.96 | 68.68 |
| reference P10, size 940 x 1600mm | nr | 1.10 | 15.40 | 51.81 | 10.08 | 77.29 |
| reference SO6, size 1140 x 1180mm | nr | 1.10 | 15.40 | 52.04 | 10.12 | 77.56 |

| | Unit | Labour | Hours £ | Mat'ls £ | O & P £ | Total £ |
|---|---|---|---|---|---|---|
| Aluminium flashing EDP 0000 for window | | | | | | |
| reference CO2, size 550 x 780mm | nr | 0.45 | 6.30 | 45.33 | 7.74 | 59.37 |
| reference CO4, size 550 x 080mm | nr | 0.45 | 6.30 | 58.02 | 9.65 | 73.97 |
| reference FO6, size 660 x 1180mm | nr | 1.00 | 14.00 | 62.47 | 11.47 | 87.94 |
| reference MO4, size 780 x 980mm | nr | 1.00 | 14.00 | 60.94 | 11.24 | 86.18 |
| reference MO6, size 780 x 1180mm | nr | 1.00 | 14.00 | 64.02 | 11.70 | 89.72 |
| reference MO8, size 780 x 140mm | nr | 1.00 | 14.00 | 68.15 | 12.32 | 94.47 |
| reference P10, size 940 x 1600mm | nr | 1.10 | 15.40 | 72.63 | 13.20 | 101.23 |
| reference SO6, size 1140 x 1180mm | nr | 1.10 | 15.40 | 67.89 | 12.49 | 95.78 |
| Aluminium twin flashing EBL 0000 for window | | | | | | |
| reference CO2, size 550 x 780mm | nr | 0.45 | 6.30 | 75.25 | 12.23 | 93.78 |
| reference CO4, size 550 x 080mm | nr | 0.45 | 6.30 | 80.27 | 12.99 | 99.56 |
| reference FO6, size 660 x 1180mm | nr | 1.00 | 14.00 | 89.94 | 15.59 | 119.53 |
| reference MO4, size 780 x 980mm | nr | 1.00 | 14.00 | 93.04 | 16.06 | 123.10 |
| reference MO6, size 780 x 1180mm | nr | 1.00 | 14.00 | 96.44 | 16.57 | 127.01 |
| reference MO8, size 780 x 140mm | nr | 1.00 | 14.00 | 102.33 | 17.45 | 133.78 |

| | Unit | Labour | Hours £ | Mat'ls £ | O & P £ | Total £ |
|---|---|---|---|---|---|---|
| **Twin flashing EBL range (cont'd)** | | | | | | |
| reference P10, size 940 x 1600mm | nr | 1.10 | 15.40 | 105.37 | 18.12 | 138.89 |
| reference SO6, size 1140 x 1180mm | nr | 1.10 | 15.40 | 107.11 | 18.38 | 140.89 |
| Aluminium twin flashing EBH 0000 for window | | | | | | |
| reference CO2, size 550 x 780mm | nr | 0.45 | 6.30 | 102.07 | 16.26 | 124.63 |
| reference CO4, size 550 x 080mm | nr | 0.45 | 6.30 | 105.73 | 16.80 | 128.83 |
| reference FO6, size 660 x 1180mm | nr | 1.00 | 14.00 | 118.04 | 19.81 | 151.85 |
| reference MO4, size 780 x 980mm | nr | 1.00 | 14.00 | 123.68 | 20.65 | 158.33 |
| reference MO6, size 780 x 1180mm | nr | 1.00 | 14.00 | 129.74 | 21.56 | 165.30 |
| reference MO8, size 780 x 140mm | nr | 1.00 | 14.00 | 132.44 | 21.97 | 168.41 |
| reference P10, size 940 x 1600mm | nr | 1.10 | 15.40 | 139.88 | 23.29 | 178.57 |
| reference SO6, size 1140 x 1180mm | nr | 1.10 | 15.40 | 145.07 | 24.07 | 184.54 |
| Aluminium twin flashing EBL 0050 for window | | | | | | |
| reference CO2, size 550 x 780mm | nr | 0.45 | 6.30 | 100.38 | 16.00 | 122.68 |
| reference CO4, size 550 x 080mm | nr | 0.45 | 6.30 | 105.48 | 16.77 | 128.55 |

| | Unit | Labour | Hours £ | Mat'ls £ | O & P £ | Total £ |
|---|---|---|---|---|---|---|
| Aluminium twin flashing EBH 0050 for window | | | | | | |
| reference CO2, size 550 x 780mm | nr | 0.45 | 6.30 | 116.54 | 18.43 | 141.27 |
| reference CO4, size 550 x 080mm | nr | 0.45 | 6.30 | 121.25 | 19.13 | 146.68 |
| Aluminium coupled flashing EKL 0021 for window | | | | | | |
| reference CO2, size 550 x 780mm | nr | 0.45 | 6.30 | 85.01 | 13.70 | 105.01 |
| reference CO4, size 550 x 080mm | nr | 0.45 | 6.30 | 87.86 | 14.12 | 108.28 |
| reference FO6, size 660 x 1180mm | nr | 1.00 | 14.00 | 95.34 | 16.40 | 125.74 |
| reference MO4, size 780 x 980mm | nr | 1.00 | 14.00 | 95.34 | 16.40 | 125.74 |
| reference MO6, size 780 x 1180mm | nr | 1.00 | 14.00 | 98.47 | 16.87 | 129.34 |
| reference MO8, size 780 x 140mm | nr | 1.00 | 14.00 | 103.47 | 17.62 | 135.09 |
| reference P10, size 940 x 1600mm | nr | 1.10 | 15.40 | 112.60 | 19.20 | 147.20 |
| reference SO6, size 1140 x 1180mm | nr | 1.10 | 15.40 | 112.27 | 19.15 | 146.82 |
| Aluminium coupled flashing EKH 0021 for window | | | | | | |
| reference CO2, size 550 x 780mm | nr | 0.45 | 6.30 | 110.29 | 17.49 | 134.08 |
| reference CO4, size 550 x 080mm | nr | 0.45 | 6.30 | 113.54 | 17.98 | 137.82 |

| | Unit | Labour | Hours £ | Mat'ls £ | O & P £ | Total £ |
|---|---|---|---|---|---|---|
| **Coupled flashing EKH range (cont'd)** | | | | | | |
| reference FO6, size 660 x 1180mm | nr | 1.00 | 14.00 | 124.03 | 20.70 | 158.73 |
| reference MO4, size 780 x 980mm | nr | 1.00 | 14.00 | 124.03 | 20.70 | 158.73 |
| reference MO6, size 780 x 1180mm | nr | 1.00 | 14.00 | 127.59 | 21.24 | 162.83 |
| reference MO8, size 780 x 140mm | nr | 1.00 | 14.00 | 133.64 | 22.15 | 169.79 |
| reference P10, size 940 x 1600mm | nr | 1.10 | 15.40 | 146.88 | 24.34 | 186.62 |
| reference SO6, size 1140 x 1180mm | nr | 1.10 | 15.40 | 145.27 | 24.10 | 184.77 |
| Aluminium duo flashing EKL 0012 for window | | | | | | |
| reference CO2, size 550 x 780mm | nr | 0.45 | 6.30 | 94.76 | 15.16 | 116.22 |
| reference CO4, size 550 x 080mm | nr | 0.45 | 6.30 | 97.18 | 15.52 | 119.00 |
| reference FO6, size 660 x 1180mm | nr | 1.00 | 14.00 | 103.59 | 17.64 | 135.23 |
| reference MO4, size 780 x 980mm | nr | 1.00 | 14.00 | 103.44 | 17.62 | 135.06 |
| reference MO6, size 780 x 1180mm | nr | 1.00 | 14.00 | 103.55 | 17.63 | 135.18 |
| reference MO8, size 780 x 140mm | nr | 1.00 | 14.00 | 105.76 | 17.96 | 137.72 |
| reference P10, size 940 x 1600mm | nr | 1.10 | 15.40 | 114.83 | 19.53 | 149.76 |
| reference SO6, size 1140 x 1180mm | nr | 1.10 | 15.40 | 110.61 | 18.90 | 144.91 |

| | Unit | Labour | Hours £ | Mat'ls £ | O & P £ | Total £ |
|---|---|---|---|---|---|---|
| Aluminium duo flashing EKH 0012 for window | | | | | | |
| reference CO2, size 550 x 780mm | nr | 0.45 | 6.30 | 104.04 | 16.55 | 126.89 |
| reference CO4, size 550 x 080mm | nr | 0.45 | 6.3,0 | 105.32 | 16.74 | 128.36 |
| reference FO6, size 660 x 1180mm | nr | 1.00 | 14.00 | 112.61 | 18.99 | 145.60 |
| reference MO4, size 780 x 980mm | nr | 1.00 | 14.00 | 113.88 | 19.18 | 147.06 |
| reference MO6, size 780 x 1180mm | nr | 1.00 | 14.00 | 116.55 | 19.58 | 150.13 |
| reference MO8, size 780 x 140mm | nr | 1.00 | 14.00 | 118.83 | 19.92 | 152.75 |
| reference P10, size 940 x 1600mm | nr | 1.10 | 15.40 | 127.46 | 21.43 | 164.29 |
| reference SO6, size 1140 x 1180mm | nr | 1.10 | 15.40 | 126.87 | 21.34 | 163.61 |
| Aluminium quatro flashing EKL 0022 for window | | | | | | |
| reference CO2, size 550 x 780mm | nr | 0.45 | 6.30 | 218.84 | 33.77 | 258.91 |
| reference CO4, size 550 x 080mm | nr | 0.45 | 6.30 | 225.34 | 34.75 | 266.39 |
| reference FO6, size 660 x 1180mm | nr | 1.00 | 14.00 | 237.14 | 37.67 | 288.81 |
| reference MO4, size 780 x 980mm | nr | 1.00 | 14.00 | 240.11 | 38.12 | 292.23 |
| reference MO6, size 780 x 1180mm | nr | 1.00 | 14.00 | 245.34 | 38.90 | 298.24 |
| reference MO8, size 780 x 140mm | nr | 1.00 | 14.00 | 252.31 | 39.95 | 306.26 |

| | Unit | Labour | Hours £ | Mat'ls £ | O & P £ | Total £ |
|---|---|---|---|---|---|---|
| **Quatro flashing EKL range (cont'd)** | | | | | | |
| reference P10, size 940 x 1600mm | nr | 1.10 | 15.40 | 271.25 | 43.00 | 329.65 |
| reference SO6, size 1140 x 1180mm | nr | 1.10 | 15.40 | 267.53 | 42.44 | 325.37 |
| Aluminium quatro flashing EKL 0022 for window | | | | | | |
| reference CO2, size 550 x 780mm | nr | 0.45 | 6.30 | 242.31 | 37.29 | 285.90 |
| reference CO4, size 550 x 080mm | nr | 0.45 | 6.30 | 247.61 | 38.09 | 292.00 |
| reference FO6, size 660 x 1180mm | nr | 1.00 | 14.00 | 262.43 | 41.46 | 317.89 |
| reference MO4, size 780 x 980mm | nr | 1.00 | 14.00 | 265.94 | 41.99 | 321.93 |
| reference MO6, size 780 x 1180mm | nr | 1.00 | 14.00 | 270.34 | 42.65 | 326.99 |
| reference MO8, size 780 x 140mm | nr | 1.00 | 14.00 | 279.75 | 44.06 | 337.81 |
| reference P10, size 940 x 1600mm | nr | 1.10 | 15.40 | 300.04 | 47.32 | 362.76 |
| reference SO6, size 1140 x 1180mm | nr | 1.10 | 15.40 | 297.48 | 46.93 | 359.81 |

# Part Two

## PROJECT COSTS

| | Qty | Unit | Labour | Hours £ | Mat'ls £ | O & P £ | Total £ |
|---|---|---|---|---|---|---|---|
| **CLAY/CONCRETE ROOF TILING** | | | | | | | |
| **Marley Marlden tiles** | | | | | | | |
| **Two-bedroom bungalow** | | | | | | | |
| Marley Marlden Plain clay tiles size 267 × 168mm on felt and battens | 180 | m2 | 259 | 3,626 | 6,076 | 1,455 | 11,157 |
| Extra for | | | | | | | |
| cloak verge system | 20 | m | 5 | 70 | 219 | 43 | 332 |
| double course at eaves | 36 | m | 32 | 448 | 111 | 84 | 643 |
| segmental ridge tile | 18 | m | 11 | 154 | 250 | 61 | 465 |
| ridge vent terminal | 1 | nr | 1 | 14 | 38 | 8 | 60 |
| | | **Carried to summary** | | | | | 12,657 |
| **Three-bedroom house with one dormer window** | | | | | | | |
| Marley Marlden Plain clay tiles size 267 × 168mm on felt and battens | 350 | m2 | 504 | 7,056 | 11,816 | 2,831 | 21,703 |
| Extra for | | | | | | | |
| double course at eaves | 76 | m | 68 | 952 | 235 | 178 | 1,365 |
| segmental ridge tile | 8 | m | 5 | 70 | 111 | 27 | 208 |
| | | | | **Carried forward** | | | 23,276 |

| | Qty | Unit | Labour | Hours £ | Mat'ls £ | O & P £ | Total £ |
|---|---|---|---|---|---|---|---|
| | | | | **Brought forward** | | | 23,276 |
| **Three-bedroom house (cont'd)** | | | | | | | |
| bonnet hip tile | 48 | m | 29 | 406 | 347 | 113 | 866 |
| ridge vent terminal | 1 | nr | 1 | 14 | 38 | 8 | 60 |
| gas vent terminal | 1 | nr | 1 | 14 | 57 | 11 | 82 |
| lead apron | 3 | m | 3 | 42 | 22 | 10 | 74 |
| lead flashing | 6 | m | 5 | 70 | 35 | 16 | 121 |
| lead sill | 3 | m | 3 | 42 | 22 | 10 | 74 |
| | | | **Carried to summary** | | | | 24,492 |
| **Five-bedroom house with two dormer windows** | | | | | | | |
| Marley Marlden Plain clay tiles size 267 × 168mm on felt and battens | 500 | m2 | 720 | 10,080 | 16,880 | 4,044 | 31,004 |
| Extra for | | | | | | | |
| double course at eaves | 90 | m | 81 | 1,134 | 279 | 212 | 1,625 |
| cloak verge system | 60 | m | 15 | 210 | 657 | 130 | 997 |
| segmental ridge tile | 25 | m | 15 | 210 | 347 | 84 | 641 |
| ridge vent terminal | 1 | nr | 1 | 14 | 38 | 8 | 60 |
| gas vent terminal | 1 | nr | 1 | 14 | 57 | 11 | 82 |
| lead apron | 6 | m | 6 | 84 | 44 | 19 | 147 |
| lead flashing | 12 | m | 10 | 140 | 70 | 32 | 242 |
| lead sill | 6 | m | 6 | 84 | 44 | 19 | 147 |
| | | | **Carried to summary** | | | | 34,944 |

| | Qty | Unit | Labour | Hours £ | Mat'ls £ | O & P £ | Total £ |
|---|---|---|---|---|---|---|---|
| **Marley Thaxton tiles** | | | | | | | |
| **Two-bedroom bungalow** | | | | | | | |
| Marley Thaxton clay tiles size 270 × 168mm on felt and battens | 180 | m2 | 259 | 3,626 | 6,163 | 1,468 | 11,257 |
| Extra for | | | | | | | |
| cloak verge system | 20 | m | 5 | 70 | 219 | 43 | 332 |
| double course at eaves | 36 | m | 32 | 448 | 111 | 84 | 643 |
| segmental ridge tile | 18 | m | 11 | 154 | 60 | 32 | 246 |
| ridge vent terminal | 1 | nr | 1 | 14 | 38 | 8 | 60 |
| | | **Carried to summary** | | | | | 12,538 |
| **Three-bedroom house with one dormer window** | | | | | | | |
| Marley Thaxton clay tiles size 270 × 168mm on felt and battens | 350 | m2 | 504 | 7,056 | 11,984 | 2,856 | 21,896 |
| Extra for | | | | | | | |
| double course at eaves | 76 | m | 68 | 952 | 235 | 178 | 1,365 |
| segmental ridge tile | 8 | m | 5 | 70 | 111 | 27 | 208 |
| bonnet hip tile | 48 | m | 29 | 406 | 347 | 113 | 866 |
| ridge vent terminal | 1 | nr | 1 | 14 | 38 | 8 | 60 |
| gas vent terminal | 1 | nr | 1 | 14 | 57 | 11 | 82 |
| lead apron | 3 | m | 3 | 42 | 22 | 10 | 74 |
| | | | | **Carried forward** | | | 24,550 |

| | Qty | Unit | Labour | Hours £ | Mat'ls £ | O & P £ | Total £ |
|---|---|---|---|---|---|---|---|
| | | | **Brought forward** | | | | 24,550 |
| lead flashing | 6 | m | 5 | 70 | 35 | 16 | 121 |
| lead sill | 3 | m | 3 | 42 | 22 | 10 | 74 |
| | | | **Carried to summary** | | | | 24,744 |

**Five-bedroom house with two dormer windows**

| | Qty | Unit | Labour | Hours £ | Mat'ls £ | O & P £ | Total £ |
|---|---|---|---|---|---|---|---|
| Marley Thaxton Plain clay tiles size 267 × 168mm on felt and battens | 500 | m2 | 720 | 10,080 | 17,120 | 4,080 | 31,280 |
| Extra for | | | | | | | |
| double course at eaves | 90 | m | 81 | 1,134 | 279 | 212 | 1,625 |
| cloak verge system | 60 | m | 15 | 210 | 657 | 130 | 997 |
| segmental ridge tile | 25 | m | 15 | 210 | 347 | 84 | 641 |
| ridge vent terminal | 1 | nr | 1 | 14 | 38 | 8 | 60 |
| gas vent terminal | 1 | nr | 1 | 14 | 57 | 11 | 82 |
| lead apron | 6 | m | 6 | 84 | 44 | 19 | 147 |
| lead flashing | 12 | m | 10 | 140 | 70 | 32 | 242 |
| lead sill | 6 | m | 6 | 84 | 44 | 19 | 147 |
| | | | **Carried to summary** | | | | 35,220 |

| | Qty | Unit | Labour | Hours £ | Mat'ls £ | O & P £ | Total £ |
|---|---|---|---|---|---|---|---|
| **Marley Heritage tiles** | | | | | | | |
| **Two-bedroom bungalow** | | | | | | | |
| Marley Heritage clay tiles size 267 × 168mm on felt and battens | 180 | m2 | 259 | 3,626 | 6,210 | 1,475 | 11,311 |
| Extra for | | | | | | | |
| cloak verge system | 20 | m | 5 | 70 | 219 | 43 | 332 |
| double course at eaves | 36 | m | 32 | 448 | 111 | 84 | 643 |
| segmental ridge tile | 18 | m | 11 | 154 | 60 | 32 | 246 |
| ridge vent terminal | 1 | nr | 1 | 14 | 38 | 8 | 60 |
| | | **Carried to summary** | | | | | 12,593 |
| **Three-bedroom house with one dormer window** | | | | | | | |
| Marley Heritage clay tiles size 267 × 168mm on felt and battens | 350 | m2 | 504 | 7,056 | 12,075 | 2,870 | 22,001 |
| Extra for | | | | | | | |
| double course at eaves | 76 | m | 52 | 728 | 235 | 144 | 1,107 |
| segmental ridge tile | 8 | m | 5 | 70 | 111 | 27 | 208 |
| bonnet hip tile | 48 | m | 29 | 406 | 347 | 113 | 866 |
| ridge vent terminal | 1 | nr | 1 | 14 | 38 | 8 | 60 |
| gas vent terminal | 1 | nr | 1 | 14 | 57 | 11 | 82 |
| lead apron | 3 | m | 3 | 42 | 22 | 10 | 74 |
| | | | | **Carried forward** | | | 24,397 |

| | Qty | Unit | Labour | Hours £ | Mat'ls £ | O & P £ | Total £ |
|---|---|---|---|---|---|---|---|
| | | | | **Brought forward** | | | 24,397 |
| lead flashing | 6 | m | 5 | 70 | 35 | 16 | 121 |
| lead sill | 3 | m | 3 | 42 | 22 | 10 | 74 |
| | | | **Carried to summary** | | | | 24,591 |

**Five-bedroom house with two dormer windows**

| | Qty | Unit | Labour | Hours £ | Mat'ls £ | O & P £ | Total £ |
|---|---|---|---|---|---|---|---|
| Marley Heritage clay tiles size 267 × 168mm on felt and battens | 500 | m2 | 720 | 10,080 | 17,250 | 4,100 | 31,430 |
| Extra for | | | | | | | |
| double course at eaves | 90 | m | 81 | 1,134 | 279 | 212 | 1,625 |
| cloak verge system | 60 | m | 15 | 210 | 657 | 130 | 997 |
| segmental ridge tile | 25 | m | 15 | 210 | 347 | 84 | 641 |
| ridge vent terminal | 1 | nr | 1 | 14 | 38 | 8 | 60 |
| gas vent terminal | 1 | nr | 1 | 14 | 57 | 11 | 82 |
| lead apron | 6 | m | 6 | 84 | 44 | 19 | 147 |
| lead flashing | 12 | m | 10 | 140 | 70 | 32 | 242 |
| lead sill | 6 | m | 6 | 84 | 44 | 19 | 147 |
| | | | **Carried to summary** | | | | 35,369 |

| | Qty | Unit | Labour | Hours £ | Mat'ls £ | O & P £ | Total £ |
|---|---|---|---|---|---|---|---|
| **Marley Plain tiles** | | | | | | | |
| **Two-bedroom bungalow** | | | | | | | |
| Marley Plain clay tiles size 267 × 168mm on felt and battens | 180 | m2 | 259 | 3,626 | 6,012 | 1,446 | 11,084 |
| Extra for | | | | | | | |
| cloak verge system | 20 | m | 5 | 70 | 219 | 43 | 332 |
| double course at eaves | 36 | m | 32 | 448 | 111 | 84 | 643 |
| segmental ridge tile | 18 | m | 11 | 154 | 60 | 32 | 246 |
| ridge vent terminal | 1 | nr | 1 | 14 | 38 | 8 | 60 |
| | | **Carried to summary** | | | | | 12,365 |
| **Three-bedroom house with one dormer window** | | | | | | | |
| Marley Plain clay tiles size 267 × 168mm on felt and battens | 350 | m2 | 504 | 7,056 | 11,690 | 2,812 | 21,558 |
| Extra for | | | | | | | |
| double course at eaves | 76 | m | 52 | 728 | 235 | 144 | 1,107 |
| segmental ridge tile | 8 | m | 5 | 70 | 111 | 27 | 208 |
| bonnet hip tile | 48 | m | 29 | 406 | 347 | 113 | 866 |
| ridge vent terminal | 1 | nr | 1 | 14 | 38 | 8 | 60 |
| gas vent terminal | 1 | nr | 1 | 14 | 57 | 11 | 82 |
| lead apron | 3 | m | 3 | 42 | 22 | 10 | 74 |
| | | | | **Carried forward** | | | 23,955 |

| | Qty | Unit | Labour | Hours £ | Mat'ls £ | O & P £ | Total £ |
|---|---|---|---|---|---|---|---|
| | | | | **Brought forward** | | | 23,955 |
| lead flashing | 6 | m | 5 | 70 | 35 | 16 | 121 |
| lead sill | 3 | m | 3 | 42 | 22 | 10 | 74 |
| | | **Carried to summary** | | | | | 24,149 |

**Five-bedroom house with two dormer windows**

| | Qty | Unit | Labour | Hours £ | Mat'ls £ | O & P £ | Total £ |
|---|---|---|---|---|---|---|---|
| Marley Plain clay tiles size 267 × 168mm on felt and battens | 500 | m2 | 720 | 10,080 | 16,700 | 4,017 | 30,797 |
| Extra for | | | | | | | |
| double course at eaves | 90 | m | 81 | 1,134 | 279 | 212 | 1,625 |
| cloak verge system | 60 | m | 15 | 210 | 657 | 130 | 997 |
| segmental ridge tile | 25 | m | 15 | 210 | 347 | 84 | 641 |
| ridge vent terminal | 1 | nr | 1 | 14 | 38 | 8 | 60 |
| gas vent terminal | 1 | nr | 1 | 14 | 57 | 11 | 82 |
| lead apron | 6 | m | 6 | 84 | 44 | 19 | 147 |
| lead flashing | 12 | m | 10 | 140 | 70 | 32 | 242 |
| lead sill | 6 | m | 6 | 84 | 44 | 19 | 147 |
| | | **Carried to summary** | | | | | 34,737 |

| | Qty | Unit | Labour | Hours £ | Mat'ls £ | O & P £ | Total £ |
|---|---|---|---|---|---|---|---|
| **Marley Modern tiles** | | | | | | | |
| **Two-bedroom bungalow** | | | | | | | |
| Marley Modern concrete tiles size 420 × 330mm on felt and battens | 180 | m2 | 106 | 1,484 | 1,884 | 505 | 3,873 |
| Extra for | | | | | | | |
| dry verge system | 20 | m | 12 | 168 | 97 | 40 | 305 |
| eaves vent system | 36 | m | 22 | 308 | 465 | 116 | 889 |
| segmental ridge tile | 18 | m | 11 | 154 | 60 | 32 | 246 |
| ridge vent terminal | 1 | nr | 1 | 14 | 38 | 8 | 60 |
| gas vent terminal | 1 | nr | 1 | 14 | 57 | 11 | 82 |
| | | **Carried to summary** | | | | | 5,454 |
| **Three-bedroom house with one dormer window** | | | | | | | |
| Marley Modern concrete tiles size 420 × 330mm on felt and battens | 350 | m2 | 206 | 2,884 | 3,664 | 982 | 7,530 |
| Extra for | | | | | | | |
| eaves vent system | 76 | m | 46 | 644 | 983 | 244 | 1,871 |
| Modern ridge tile | 8 | m | 3 | 42 | 59 | 15 | 116 |
| 1/3 hip tile | 48 | m | 19 | 266 | 348 | 92 | 706 |
| ridge vent terminal | 1 | nr | 1 | 14 | 38 | 8 | 60 |
| gas vent terminal | 1 | nr | 1 | 14 | 57 | 11 | 82 |
| lead apron | 3 | m | 3 | 42 | 22 | 10 | 74 |
| | | | | **Carried forward** | | | 10,439 |

| | Qty | Unit | Labour | Hours £ | Mat'ls £ | O & P £ | Total £ |
|---|---|---|---|---|---|---|---|
| | | | | **Brought forward** | | | 10,439 |
| lead flashing | 6 | m | 5 | 70 | 35 | 16 | 121 |
| lead sill | 3 | m | 3 | 42 | 22 | 10 | 74 |
| | | | **Carried to summary** | | | | 10,633 |

**Five-bedroom house with two dormer windows**

| | Qty | Unit | Labour | Hours £ | Mat'ls £ | O & P £ | Total £ |
|---|---|---|---|---|---|---|---|
| Marley Modern concrete tiles size 420 × 330mm on felt and battens | 500 | m2 | 295 | 4,130 | 5,235 | 1,405 | 10,770 |
| Extra for | | | | | | | |
| dry verge system | 60 | m | 36 | 504 | 291 | 119 | 914 |
| eaves vent system | 90 | m | 54 | 756 | 288 | 157 | 1,201 |
| Modern ridge tile | 25 | m | 10 | 140 | 184 | 49 | 373 |
| ridge vent terminal | 1 | nr | 1 | 14 | 38 | 8 | 60 |
| gas vent terminal | 1 | nr | 1 | 14 | 57 | 11 | 82 |
| lead apron | 6 | m | 6 | 84 | 44 | 19 | 147 |
| lead flashing | 12 | m | 10 | 140 | 70 | 32 | 242 |
| lead sill | 6 | m | 6 | 84 | 44 | 19 | 147 |
| | | | **Carried to summary** | | | | 13,434 |

| | Qty | Unit | Labour | Hours £ | Mat'ls £ | O & P £ | Total £ |
|---|---|---|---|---|---|---|---|
| **Marley Ludlow Major tiles** | | | | | | | |
| **Two-bedroom bungalow** | | | | | | | |
| Marley Ludlow Major concrete tiles size 420 × 330mm on felt and battens | 180 | m2 | 106 | 1,484 | 1,900 | 508 | 3,892 |
| Extra for | | | | | | | |
| dry verge system | 20 | m | 12 | 168 | 97 | 40 | 305 |
| eaves vent system | 36 | m | 22 | 308 | 465 | 116 | 889 |
| segmental ridge tile | 18 | m | 11 | 154 | 60 | 32 | 246 |
| ridge vent terminal | 1 | nr | 1 | 14 | 38 | 8 | 60 |
| gas vent terminal | 1 | nr | 1 | 14 | 57 | 11 | 82 |
| | | **Carried to summary** | | | | | 5,473 |
| **Three-bedroom house with one dormer window** | | | | | | | |
| Marley Ludlow Major concrete tiles size 420 × 330mm on felt and battens | 350 | m2 | 206 | 2,884 | 3,696 | 987 | 7,567 |
| Extra for | | | | | | | |
| eaves vent system | 76 | m | 46 | 644 | 983 | 244 | 1,871 |
| Modern ridge tile | 8 | m | 3 | 42 | 59 | 15 | 116 |
| 1/3 hip tile | 48 | m | 19 | 266 | 348 | 92 | 706 |
| ridge vent terminal | 1 | nr | 1 | 14 | 38 | 8 | 60 |
| | | | | **Carried forward** | | | 10,475 |

| | Qty | Unit | Labour | Hours £ | Mat'ls £ | O & P £ | Total £ |
|---|---|---|---|---|---|---|---|
| | | | | **Brought forward** | | | 10,475 |
| gas vent terminal | 1 | nr | 1 | 14 | 57 | 11 | 82 |
| lead apron | 3 | m | 3 | 42 | 22 | 10 | 74 |
| lead flashing | 6 | m | 5 | 70 | 35 | 16 | 121 |
| lead sill | 3 | m | 3 | 42 | 22 | 10 | 74 |
| | | | **Carried to summary** | | | | 10,825 |
| **Five-bedroom house with two dormer windows** | | | | | | | |
| Marley Ludlow Major concrete tiles size 420 × 330mm on felt and battens | 500 | m2 | 295 | 4,130 | 5,280 | 1,412 | 10,822 |
| Extra for | | | | | | | |
| dry verge system | 60 | m | 36 | 504 | 291 | 119 | 914 |
| eaves vent system | 90 | m | 54 | 756 | 288 | 157 | 1,201 |
| Modern ridge tile | 25 | m | 10 | 140 | 184 | 49 | 373 |
| ridge vent terminal | 1 | nr | 1 | 14 | 38 | 8 | 60 |
| gas vent terminal | 1 | nr | 1 | 14 | 57 | 11 | 82 |
| lead apron | 6 | m | 6 | 84 | 44 | 19 | 147 |
| lead flashing | 12 | m | 10 | 140 | 70 | 32 | 242 |
| lead sill | 6 | m | 6 | 84 | 44 | 19 | 147 |
| | | | **Carried to summary** | | | | 13,485 |

| | Qty | Unit | Labour | Hours £ | Mat'ls £ | O & P £ | Total £ |
|---|---|---|---|---|---|---|---|
| **Marley Ludlow Plus tiles** | | | | | | | |
| **Two-bedroom bungalow** | | | | | | | |
| Marley Ludlow Plus concrete tiles size 420 × 330mm on felt and battens | 180 | m2 | 124 | 1,736 | 2,116 | 578 | 4,430 |
| Extra for | | | | | | | |
| dry verge system | 20 | m | 12 | 168 | 97 | 40 | 305 |
| eaves vent system | 36 | m | 22 | 308 | 465 | 116 | 889 |
| dry ridge system | 18 | m | 11 | 154 | 233 | 58 | 445 |
| ridge vent terminal | 1 | nr | 1 | 14 | 38 | 8 | 60 |
| gas vent terminal | 1 | nr | 1 | 14 | 57 | 11 | 82 |
| | | **Carried to summary** | | | | | 6,210 |
| **Three-bedroom house with one dormer window** | | | | | | | |
| Marley Ludlow Plus concrete tiles size 420 × 330mm on felt and battens | 350 | m2 | 241 | 3,374 | 4,116 | 1,124 | 8,614 |
| Extra for | | | | | | | |
| eaves vent system | 76 | m | 46 | 644 | 983 | 244 | 1,871 |
| dry ridge system | 8 | m | 5 | 70 | 104 | 26 | 200 |
| 1/3 hip tile | 48 | m | 19 | 266 | 348 | 92 | 706 |
| ridge vent terminal | 1 | nr | 1 | 14 | 38 | 8 | 60 |
| | | | | **Carried forward** | | | 11,451 |

| | Qty | Unit | Labour | Hours £ | Mat'ls £ | O & P £ | Total £ |
|---|---|---|---|---|---|---|---|
| | | | | **Brought forward** | | | 11,451 |
| gas vent terminal | 1 | nr | 1 | 14 | 57 | 11 | 82 |
| lead apron | 3 | m | 3 | 42 | 22 | 10 | 74 |
| lead flashing | 6 | m | 5 | 70 | 35 | 16 | 121 |
| lead sill | 3 | m | 3 | 42 | 22 | 10 | 74 |
| | | | **Carried to summary** | | | | 11,800 |

**Five-bedroom house with two dormer windows**

| | Qty | Unit | Labour | Hours £ | Mat'ls £ | O & P £ | Total £ |
|---|---|---|---|---|---|---|---|
| Marley Ludlow Plus concrete tiles size 420 × 330mm on felt and battens | 500 | m2 | 345 | 4,830 | 5,880 | 1,607 | 12,317 |
| Extra for | | | | | | | |
| dry verge system | 60 | m | 36 | 504 | 291 | 119 | 914 |
| eaves vent system | 90 | m | 54 | 756 | 288 | 157 | 1,201 |
| dry ridge system | 25 | m | 15 | 210 | 323 | 80 | 613 |
| ridge vent terminal | 1 | nr | 1 | 14 | 38 | 8 | 60 |
| gas vent terminal | 1 | nr | 1 | 14 | 57 | 11 | 82 |
| lead apron | 6 | m | 6 | 84 | 44 | 19 | 147 |
| lead flashing | 12 | m | 10 | 140 | 70 | 32 | 242 |
| lead sill | 6 | m | 6 | 84 | 44 | 19 | 147 |
| | | | **Carried to summary** | | | | 15,722 |

| | Qty | Unit | Labour | Hours £ | Mat'ls £ | O & P £ | Total £ |
|---|---|---|---|---|---|---|---|
| **Marley Anglia Plus tiles** | | | | | | | |
| **Two-bedroom bungalow** | | | | | | | |
| Marley Anglia Plus concrete tiles size 387 × 229mm on felt and battens | 180 | m2 | 124 | 1,736 | 2,514 | 638 | 4,888 |
| Extra for | | | | | | | |
| dry verge system | 20 | m | 12 | 168 | 97 | 40 | 305 |
| eaves vent system | 36 | m | 22 | 308 | 465 | 116 | 889 |
| dry ridge system | 18 | m | 11 | 154 | 233 | 58 | 445 |
| ridge vent terminal | 1 | nr | 1 | 14 | 38 | 8 | 60 |
| gas vent terminal | 1 | nr | 1 | 14 | 57 | 11 | 82 |
| | | **Carried to summary** | | | | | 6,668 |
| **Three-bedroom house with one dormer window** | | | | | | | |
| Marley Anglia Plus concrete tiles size 387 × 229mm on felt and battens | 350 | m2 | 241 | 3,374 | 4,889 | 1,239 | 9,502 |
| Extra for | | | | | | | |
| eaves vent system | 76 | m | 46 | 644 | 983 | 244 | 1,871 |
| dry ridge system | 8 | m | 5 | 70 | 104 | 26 | 200 |
| 1/3 hip tile | 48 | m | 19 | 266 | 348 | 92 | 706 |
| ridge vent terminal | 1 | nr | 1 | 14 | 38 | 8 | 60 |
| | | | | **Carried forward** | | | 12,340 |

| | Qty | Unit | Labour | Hours £ | Mat'ls £ | O & P £ | Total £ |
|---|---|---|---|---|---|---|---|
| | | | | Brought forward | | | 12,340 |
| gas vent terminal | 1 | nr | 1 | 14 | 57 | 11 | 82 |
| lead apron | 3 | m | 3 | 42 | 22 | 10 | 74 |
| lead flashing | 6 | m | 5 | 70 | 35 | 16 | 121 |
| lead sill | 3 | m | 3 | 42 | 22 | 10 | 74 |
| | | | Carried to summary | | | | 12,690 |

**Five-bedroom house with two dormer windows**

| | Qty | Unit | Labour | Hours £ | Mat'ls £ | O & P £ | Total £ |
|---|---|---|---|---|---|---|---|
| Marley Anglia Plus concrete tiles size 387 × 229mm on felt and battens | 500 | m2 | 345 | 4,830 | 6,985 | 1,772 | 13,587 |
| Extra for | | | | | | | |
| dry verge system | 60 | m | 36 | 504 | 291 | 119 | 914 |
| eaves vent system | 90 | m | 54 | 756 | 288 | 157 | 1,201 |
| dry ridge system | 25 | m | 15 | 210 | 323 | 80 | 613 |
| ridge vent terminal | 1 | nr | 1 | 14 | 38 | 8 | 60 |
| gas vent terminal | 1 | nr | 1 | 14 | 57 | 11 | 82 |
| lead apron | 6 | m | 6 | 84 | 44 | 19 | 147 |
| lead flashing | 12 | m | 10 | 140 | 70 | 32 | 242 |
| lead sill | 6 | m | 6 | 84 | 44 | 19 | 147 |
| | | | Carried to summary | | | | 16,992 |

| | Qty | Unit | Labour | Hours £ | Mat'ls £ | O & P £ | Total £ |
|---|---|---|---|---|---|---|---|
| **Marley Mendip tiles** | | | | | | | |
| **Two-bedroom bungalow** | | | | | | | |
| Marley Mendip concrete tiles size 420 × 330mm on felt and battens | 180 | m2 | 106 | 1,484 | 1,933 | 513 | 3,930 |
| Extra for | | | | | | | |
| dry verge system | 20 | m | 12 | 168 | 97 | 40 | 305 |
| eaves vent system | 36 | m | 22 | 308 | 465 | 116 | 889 |
| dry ridge system | 18 | m | 11 | 154 | 233 | 58 | 445 |
| ridge vent terminal | 1 | nr | 1 | 14 | 38 | 8 | 60 |
| gas vent terminal | 1 | nr | 1 | 14 | 57 | 11 | 82 |
| | | **Carried to summary** | | | | | 5,710 |
| **Three-bedroom house with one dormer window** | | | | | | | |
| Marley Mendip concrete tiles size 420 × 330mm on felt and battens | 350 | m2 | 206 | 2,884 | 3,579 | 969 | 7,432 |
| Extra for | | | | | | | |
| eaves vent system | 76 | m | 46 | 644 | 983 | 244 | 1,871 |
| dry ridge system | 8 | m | 5 | 70 | 104 | 26 | 200 |
| 1/3 hip tile | 48 | m | 19 | 266 | 348 | 92 | 706 |
| ridge vent terminal | 1 | nr | 1 | 14 | 38 | 8 | 60 |
| gas vent terminal | 1 | nr | 1 | 14 | 57 | 11 | 82 |
| | | | | **Carried forward** | | | 10,351 |

| | Qty | Unit | Labour | Hours £ | Mat'ls £ | O & P £ | Total £ |
|---|---|---|---|---|---|---|---|
| | | | | **Brought forward** | | | 10,351 |
| lead apron | 3 | m | 3 | 42 | 22 | 10 | 74 |
| lead flashing | 6 | m | 5 | 70 | 35 | 16 | 121 |
| lead sill | 3 | m | 3 | 42 | 22 | 10 | 74 |
| | | | **Carried to summary** | | | | 10,619 |
| **Five-bedroom house with two dormer windows** | | | | | | | |
| Marley Mendip concrete tiles size 420 × 330mm on felt and battens | 500 | m2 | 295 | 4,130 | 5,370 | 1,425 | 10,925 |
| Extra for | | | | | | | |
| dry verge system | 60 | m | 36 | 504 | 291 | 119 | 914 |
| eaves vent system | 90 | m | 54 | 756 | 288 | 157 | 1,201 |
| dry ridge system | 25 | m | 15 | 210 | 323 | 80 | 613 |
| ridge vent terminal | 1 | nr | 1 | 14 | 38 | 8 | 60 |
| gas vent terminal | 1 | nr | 1 | 14 | 57 | 11 | 82 |
| lead apron | 6 | m | 6 | 84 | 44 | 19 | 147 |
| lead flashing | 12 | m | 10 | 140 | 70 | 32 | 242 |
| lead sill | 6 | m | 6 | 84 | 44 | 19 | 147 |
| | | | **Carried to summary** | | | | 14,330 |

| | Qty | Unit | Labour | Hours £ | Mat'ls £ | O & P £ | Total £ |
|---|---|---|---|---|---|---|---|
| **Marley Malvern tiles** | | | | | | | |
| **Two-bedroom bungalow** | | | | | | | |
| Marley Malvern concrete tiles size 420 × 330mm on felt and battens | 180 | m2 | 106 | 1,484 | 1,863 | 502 | 3,849 |
| Extra for | | | | | | | |
| dry verge system | 20 | m | 12 | 168 | 97 | 40 | 305 |
| eaves vent system | 36 | m | 22 | 308 | 465 | 116 | 889 |
| dry ridge system | 18 | m | 11 | 154 | 233 | 58 | 445 |
| ridge vent terminal | 1 | nr | 1 | 14 | 38 | 8 | 60 |
| gas vent terminal | 1 | nr | 1 | 14 | 57 | 11 | 82 |
| | | **Carried to summary** | | | | | 5,629 |
| **Three-bedroom house with one dormer window** | | | | | | | |
| Marley Malvern concrete tiles size 420 × 330mm on felt and battens | 350 | m2 | 206 | 2,884 | 3,622 | 976 | 7,482 |
| Extra for | | | | | | | |
| eaves vent system | 76 | m | 46 | 644 | 983 | 244 | 1,871 |
| dry ridge system | 8 | m | 5 | 70 | 104 | 26 | 200 |
| 1/3 hip tile | 48 | m | 19 | 266 | 348 | 92 | 706 |
| ridge vent terminal | 1 | nr | 1 | 14 | 38 | 8 | 60 |
| gas vent terminal | 1 | nr | 1 | 14 | 57 | 11 | 82 |
| | | | | **Carried forward** | | | 10,401 |

| | Qty | Unit | Labour | Hours £ | Mat'ls £ | O & P £ | Total £ |
|---|---|---|---|---|---|---|---|
| | | | | **Brought forward** | | | 10,401 |
| lead apron | 3 | m | 3 | 42 | 22 | 10 | 74 |
| lead flashing | 6 | m | 5 | 70 | 35 | 16 | 121 |
| lead sill | 3 | m | 3 | 42 | 22 | 10 | 74 |
| | | | **Carried to summary** | | | | 10,669 |

**Five-bedroom house with two dormer windows**

| | Qty | Unit | Labour | Hours £ | Mat'ls £ | O & P £ | Total £ |
|---|---|---|---|---|---|---|---|
| Marley Malvern concrete tiles size 420 × 330mm on felt and battens | 500 | m2 | 295 | 4,130 | 5,175 | 1,396 | 10,701 |
| Extra for | | | | | | | |
| dry verge system | 60 | m | 36 | 504 | 291 | 119 | 914 |
| eaves vent system | 90 | m | 54 | 756 | 288 | 157 | 1,201 |
| dry ridge system | 25 | m | 15 | 210 | 323 | 80 | 613 |
| ridge vent terminal | 1 | nr | 1 | 14 | 38 | 8 | 60 |
| gas vent terminal | 1 | nr | 1 | 14 | 57 | 11 | 82 |
| lead apron | 6 | m | 6 | 84 | 44 | 19 | 147 |
| lead flashing | 12 | m | 10 | 140 | 70 | 32 | 242 |
| lead sill | 6 | m | 6 | 84 | 44 | 19 | 147 |
| | | | **Carried to summary** | | | | 14,106 |

| | Qty | Unit | Labour | Hours £ | Mat'ls £ | O & P £ | Total £ |
|---|---|---|---|---|---|---|---|
| **Marley Double Roman tiles** | | | | | | | |
| **Two-bedroom bungalow** | | | | | | | |
| Marley Double Roman concrete tiles size 420 × 330mm on felt and battens | 180 | m2 | 106 | 1,484 | 1,942 | 514 | 3,940 |
| Extra for | | | | | | | |
| dry verge system | 20 | m | 12 | 168 | 97 | 40 | 305 |
| eaves vent system | 36 | m | 22 | 308 | 465 | 116 | 889 |
| dry ridge system | 18 | m | 11 | 154 | 233 | 58 | 445 |
| ridge vent terminal | 1 | nr | 1 | 14 | 38 | 8 | 60 |
| gas vent terminal | 1 | nr | 1 | 14 | 57 | 11 | 82 |
| | | **Carried to summary** | | | | | 5,720 |
| **Three-bedroom house with one dormer window** | | | | | | | |
| Marley Double Roman concrete tiles size 420 × 330mm on felt and battens | 350 | m2 | 206 | 2,884 | 3,776 | 999 | 7,659 |
| Extra for | | | | | | | |
| eaves vent system | 76 | m | 46 | 644 | 983 | 244 | 1,871 |
| dry ridge system | 8 | m | 5 | 70 | 104 | 26 | 200 |
| 1/3 hip tile | 48 | m | 19 | 266 | 348 | 92 | 706 |
| | | | | **Carried forward** | | | 10,436 |

| | Qty | Unit | Labour | Hours £ | Mat'ls £ | O & P £ | Total £ |
|---|---|---|---|---|---|---|---|
| | | | | **Brought forward** | | | 10,436 |
| ridge vent terminal | 1 | nr | 1 | 14 | 38 | 8 | 60 |
| gas vent terminal | 1 | nr | 1 | 14 | 57 | 11 | 82 |
| lead apron | 3 | m | 3 | 42 | 22 | 10 | 74 |
| lead flashing | 6 | m | 5 | 70 | 35 | 16 | 121 |
| lead sill | 3 | m | 3 | 42 | 22 | 10 | 74 |
| | | **Carried to summary** | | | | | 10,846 |

**Five-bedroom house with two dormer windows**

| | Qty | Unit | Labour | Hours £ | Mat'ls £ | O & P £ | Total £ |
|---|---|---|---|---|---|---|---|
| Marley Double Roman concrete tiles size 420 × 330mm on felt and battens | 500 | m2 | 295 | 4,130 | 5,395 | 1,429 | 10,954 |
| Extra for | | | | | | | |
| dry verge system | 60 | m | 36 | 504 | 291 | 119 | 914 |
| eaves vent system | 90 | m | 54 | 756 | 288 | 157 | 1,201 |
| dry ridge system | 25 | m | 15 | 210 | 323 | 80 | 613 |
| ridge vent terminal | 1 | nr | 1 | 14 | 38 | 8 | 60 |
| gas vent terminal | 1 | nr | 1 | 14 | 57 | 11 | 82 |
| lead apron | 6 | m | 6 | 84 | 44 | 19 | 147 |
| lead flashing | 12 | m | 10 | 140 | 70 | 32 | 242 |
| lead sill | 6 | m | 6 | 84 | 44 | 19 | 147 |
| | | **Carried to summary** | | | | | 14,359 |

| | Qty | Unit | Labour | Hours £ | Mat'ls £ | O & P £ | Total £ |
|---|---|---|---|---|---|---|---|
| **Marley Bold Roll tiles** | | | | | | | |
| **Two-bedroom bungalow** | | | | | | | |
| Marley Bold Roll concrete tiles size 420 × 330mm on felt and battens | 180 | m2 | 106 | 1,484 | 1,803 | 493 | 3,780 |
| Extra for | | | | | | | |
| dry verge system | 20 | m | 12 | 168 | 97 | 40 | 305 |
| eaves vent system | 36 | m | 22 | 308 | 465 | 116 | 889 |
| dry ridge system | 18 | m | 11 | 154 | 233 | 58 | 445 |
| ridge vent terminal | 1 | nr | 1 | 14 | 38 | 8 | 60 |
| gas vent terminal | 1 | nr | 1 | 14 | 57 | 11 | 82 |
| | | **Carried to summary** | | | | | 5,560 |
| **Three-bedroom house with one dormer window** | | | | | | | |
| Marley Bold Roll concrete tiles size 420 × 330mm on felt and battens | 350 | m2 | 206 | 2,884 | 3,507 | 959 | 7,350 |
| Extra for | | | | | | | |
| eaves vent system | 76 | m | 46 | 644 | 983 | 244 | 1,871 |
| dry ridge system | 8 | m | 5 | 70 | 104 | 26 | 200 |
| 1/3 hip tile | 48 | m | 19 | 266 | 348 | 92 | 706 |
| | | | | **Carried forward** | | | 10,127 |

| | Qty | Unit | Labour | Hours £ | Mat'ls £ | O & P £ | Total £ |
|---|---|---|---|---|---|---|---|
| | | | | **Brought forward** | | | 10,127 |
| ridge vent terminal | 1 | nr | 1 | 14 | 38 | 8 | 60 |
| gas vent terminal | 1 | nr | 1 | 14 | 57 | 11 | 82 |
| lead apron | 3 | m | 3 | 42 | 22 | 10 | 74 |
| lead flashing | 6 | m | 5 | 70 | 35 | 16 | 121 |
| lead sill | 3 | m | 3 | 42 | 22 | 10 | 74 |
| | | | **Carried to summary** | | | | 10,536 |

**Five-bedroom house with two dormer windows**

| | Qty | Unit | Labour | Hours £ | Mat'ls £ | O & P £ | Total £ |
|---|---|---|---|---|---|---|---|
| Marley Bold Roll concrete tiles size 420 × 330mm on felt and battens | 500 | m2 | 295 | 4,130 | 5,010 | 1,371 | 10,511 |
| Extra for | | | | | | | |
| dry verge system | 60 | m | 36 | 504 | 291 | 119 | 914 |
| eaves vent system | 90 | m | 54 | 756 | 288 | 157 | 1,201 |
| dry ridge system | 25 | m | 15 | 210 | 323 | 80 | 613 |
| ridge vent terminal | 1 | nr | 1 | 14 | 38 | 8 | 60 |
| gas vent terminal | 1 | nr | 1 | 14 | 57 | 11 | 82 |
| lead apron | 6 | m | 6 | 84 | 44 | 19 | 147 |
| lead flashing | 12 | m | 10 | 140 | 70 | 32 | 242 |
| lead sill | 6 | m | 6 | 84 | 44 | 19 | 147 |
| | | | **Carried to summary** | | | | 13,916 |

| | Qty | Unit | Labour | Hours £ | Mat'ls £ | O & P £ | Total £ |
|---|---|---|---|---|---|---|---|
| **Marley Wessex tiles** | | | | | | | |
| **Two-bedroom bungalow** | | | | | | | |
| Marley Wessex concrete tiles size 420 × 330mm on felt and battens | 180 | m2 | 106 | 1,484 | 2,593 | 612 | 4,689 |
| Extra for | | | | | | | |
| dry verge system | 20 | m | 12 | 168 | 97 | 40 | 305 |
| eaves vent system | 36 | m | 22 | 308 | 465 | 116 | 889 |
| dry ridge system | 18 | m | 11 | 154 | 233 | 58 | 445 |
| ridge vent terminal | 1 | nr | 1 | 14 | 38 | 8 | 60 |
| gas vent terminal | 1 | nr | 1 | 14 | 57 | 11 | 82 |
| | | **Carried to summary** | | | | | 6,469 |
| **Three-bedroom house with one dormer window** | | | | | | | |
| Marley Wessex concrete tiles size 420 × 330mm on felt and battens | 350 | m2 | 206 | 2,884 | 5,043 | 1,189 | 9,116 |
| Extra for | | | | | | | |
| eaves vent system | 76 | m | 46 | 644 | 983 | 244 | 1,871 |
| dry ridge system | 8 | m | 5 | 70 | 104 | 26 | 200 |
| 1/3 hip tile | 48 | m | 19 | 266 | 348 | 92 | 706 |
| | | | | **Carried forward** | | | 11,893 |

| | Qty | Unit | Labour | Hours £ | Mat'ls £ | O & P £ | Total £ |
|---|---|---|---|---|---|---|---|
| | | | | **Brought forward** | | | 11,893 |
| ridge vent terminal | 1 | nr | 1 | 14 | 38 | 8 | 60 |
| gas vent terminal | 1 | nr | 1 | 14 | 57 | 11 | 82 |
| lead apron | 3 | m | 3 | 42 | 22 | 10 | 74 |
| lead flashing | 6 | m | 5 | 70 | 35 | 16 | 121 |
| lead sill | 3 | m | 3 | 42 | 22 | 10 | 74 |
| | | | **Carried to summary** | | | | 12,303 |
| **Five-bedroom house with two dormer windows** | | | | | | | |
| Marley Wessex concrete tiles size 420 × 330mm on felt and battens | 500 | m2 | 295 | 4,130 | 7,205 | 1,700 | 13,035 |
| Extra for | | | | | | | |
| dry verge system | 60 | m | 36 | 504 | 291 | 119 | 914 |
| eaves vent system | 90 | m | 54 | 756 | 288 | 157 | 1,201 |
| dry ridge system | 25 | m | 15 | 210 | 323 | 80 | 613 |
| ridge vent terminal | 1 | nr | 1 | 14 | 38 | 8 | 60 |
| gas vent terminal | 1 | nr | 1 | 14 | 57 | 11 | 82 |
| lead apron | 6 | m | 6 | 84 | 44 | 19 | 147 |
| lead flashing | 12 | m | 10 | 140 | 70 | 32 | 242 |
| lead sill | 6 | m | 6 | 84 | 44 | 19 | 147 |
| | | | **Carried to summary** | | | | 16,440 |

| | Qty | Unit | Labour | Hours £ | Mat'ls £ | O & P £ | Total £ |
|---|---|---|---|---|---|---|---|
| **Lafarge Norfolk tiles** | | | | | | | |
| **Two-bedroom bungalow** | | | | | | | |
| Lafarge Norfolk concrete tiles size 381 × 287mm on felt and battens | 180 | m2 | 106 | 1,484 | 2,264 | 562 | 4,310 |
| Extra for | | | | | | | |
| cloaked verge course | 20 | m | 8 | 112 | 156 | 40 | 308 |
| eaves vent system | 36 | m | 14 | 196 | 355 | 83 | 634 |
| universal ridge tile | 18 | m | 7 | 98 | 180 | 42 | 320 |
| gas flue tile | 1 | nr | 1 | 14 | 52 | 10 | 76 |
| | | **Carried to summary** | | | | | 5,339 |
| **Three-bedroom house with one dormer window** | | | | | | | |
| Lafarge Norfolk concrete tiles size 381 × 287mm on felt and battens | 350 | m2 | 206 | 2,884 | 4,403 | 1,093 | 8,380 |
| Extra for | | | | | | | |
| eaves vent system | 76 | m | 30 | 420 | 750 | 176 | 1,346 |
| universal ridge tile | 8 | m | 5 | 70 | 104 | 26 | 200 |
| 1/3 hip tile | 48 | m | 19 | 266 | 479 | 112 | 857 |
| | | | | **Carried forward** | | | 10,782 |

| | Qty | Unit | Labour | Hours £ | Mat'ls £ | O & P £ | Total £ |
|---|---|---|---|---|---|---|---|
| | | | | **Brought forward** | | | 10,782 |
| gas flue tile | 1 | nr | 1 | 14 | 52 | 10 | 76 |
| lead apron | 3 | m | 3 | 42 | 22 | 10 | 74 |
| lead flashing | 6 | m | 5 | 70 | 35 | 16 | 121 |
| lead sill | 3 | m | 3 | 42 | 22 | 10 | 74 |
| | | | **Carried to summary** | | | | 11,126 |

**Five-bedroom house with two dormer windows**

| | Qty | Unit | Labour | Hours £ | Mat'ls £ | O & P £ | Total £ |
|---|---|---|---|---|---|---|---|
| Lafarge Norfolk concrete tiles size 381 × 287mm on felt and battens | 500 | m2 | 295 | 4,130 | 6,290 | 1,563 | 11,983 |
| Extra for | | | | | | | |
| cloaked verge course | 60 | m | 24 | 336 | 468 | 121 | 925 |
| eaves vent system | 76 | m | 30 | 420 | 750 | 176 | 1,346 |
| universal ridge tile | 25 | m | 10 | 140 | 100 | 36 | 276 |
| gas vent terminal | 1 | nr | 1 | 14 | 57 | 11 | 82 |
| lead apron | 6 | m | 6 | 84 | 44 | 19 | 147 |
| lead flashing | 12 | m | 10 | 140 | 70 | 32 | 242 |
| lead sill | 6 | m | 6 | 84 | 44 | 19 | 147 |
| | | | **Carried to summary** | | | | 15,147 |

| | Qty | Unit | Labour | Hours £ | Mat'ls £ | O & P £ | Total £ |
|---|---|---|---|---|---|---|---|
| **Lafarge Renown tiles** | | | | | | | |
| **Two-bedroom bungalow** | | | | | | | |
| Lafarge Renown concrete tiles size 418 × 330mm on felt and battens | 180 | m2 | 106 | 1,484 | 1,883 | 505 | 3,872 |
| Extra for | | | | | | | |
| cloaked verge course | 20 | m | 8 | 112 | 156 | 40 | 308 |
| eaves vent system | 36 | m | 14 | 196 | 355 | 83 | 634 |
| universal ridge tile | 18 | m | 7 | 98 | 180 | 42 | 320 |
| gas flue tile | 1 | nr | 1 | 14 | 52 | 10 | 76 |
| | | **Carried to summary** | | | | | 4,901 |
| **Three-bedroom house with one dormer window** | | | | | | | |
| Lafarge Renown concrete tiles size 418 × 330mm on felt and battens | 350 | m2 | 206 | 2,884 | 3,661 | 982 | 7,527 |
| Extra for | | | | | | | |
| eaves vent system | 76 | m | 30 | 420 | 750 | 176 | 1,346 |
| universal ridge tile | 8 | m | 5 | 70 | 104 | 26 | 200 |
| 1/3 hip tile | 48 | m | 19 | 266 | 479 | 112 | 857 |
| | | | | **Carried forward** | | | 9,929 |

| | Qty | Unit | Labour | Hours £ | Mat'ls £ | O & P £ | Total £ |
|---|---|---|---|---|---|---|---|
| | | | | **Brought forward** | | | 9,929 |
| gas flue tile | 1 | nr | 1 | 14 | 52 | 10 | 76 |
| lead apron | 3 | m | 3 | 42 | 22 | 10 | 74 |
| lead flashing | 6 | m | 5 | 70 | 35 | 16 | 121 |
| lead sill | 3 | m | 3 | 42 | 22 | 10 | 74 |
| | | **Carried to summary** | | | | | 10,273 |

**Five-bedroom house with two dormer windows**

| | Qty | Unit | Labour | Hours £ | Mat'ls £ | O & P £ | Total £ |
|---|---|---|---|---|---|---|---|
| Lafarge Renown concrete tiles size 418 × 330mm on felt and battens | 500 | m2 | 295 | 4,130 | 5,230 | 1,404 | 10,764 |
| Extra for | | | | | | | |
| cloaked verge course | 60 | m | 24 | 336 | 468 | 121 | 925 |
| eaves vent system | 76 | m | 30 | 420 | 750 | 176 | 1,346 |
| universal ridge tile | 25 | m | 10 | 140 | 100 | 36 | 276 |
| gas vent terminal | 1 | nr | 1 | 14 | 57 | 11 | 82 |
| lead apron | 6 | m | 6 | 84 | 44 | 19 | 147 |
| lead flashing | 12 | m | 10 | 140 | 70 | 32 | 242 |
| lead sill | 6 | m | 6 | 84 | 44 | 19 | 147 |
| | | **Carried to summary** | | | | | 13,928 |

| | Qty | Unit | Labour | Hours £ | Mat'ls £ | O & P £ | Total £ |
|---|---|---|---|---|---|---|---|
| **Lafarge Double Roman tiles** | | | | | | | |
| **Two-bedroom bungalow** | | | | | | | |
| Lafarge Double Roman concrete tiles size 418 × 330mm on felt and battens | 180 | m2 | 106 | 1,484 | 1,883 | 505 | 3,872 |
| Extra for | | | | | | | |
| cloaked verge course | 20 | m | 8 | 112 | 156 | 40 | 308 |
| eaves vent system | 36 | m | 14 | 196 | 355 | 83 | 634 |
| universal ridge tile | 18 | m | 7 | 98 | 180 | 42 | 320 |
| gas flue tile | 1 | nr | 1 | 14 | 52 | 10 | 76 |
| | | **Carried to summary** | | | | | 4,901 |
| **Three-bedroom house with one dormer window** | | | | | | | |
| Lafarge Double Roman concrete tiles size 418 × 330mm on felt and battens | 350 | m2 | 206 | 2,884 | 3,661 | 982 | 7,527 |
| Extra for | | | | | | | |
| eaves vent system | 76 | m | 30 | 420 | 750 | 176 | 1,346 |
| universal ridge tile | 8 | m | 5 | 70 | 104 | 26 | 200 |
| 1/3 hip tile | 48 | m | 19 | 266 | 479 | 112 | 857 |
| | | | | **Carried forward** | | | 9,929 |

| | Qty | Unit | Labour | Hours £ | Mat'ls £ | O & P £ | Total £ |
|---|---|---|---|---|---|---|---|
| | | | | **Brought forward** | | | 9,929 |
| gas flue tile | 1 | nr | 1 | 14 | 52 | 10 | 76 |
| lead apron | 3 | m | 3 | 42 | 22 | 10 | 74 |
| lead flashing | 6 | m | 5 | 70 | 35 | 16 | 121 |
| lead sill | 3 | m | 3 | 42 | 22 | 10 | 74 |
| | | | **Carried to summary** | | | | 10,273 |

**Five-bedroom house with two dormer windows**

| | Qty | Unit | Labour | Hours £ | Mat'ls £ | O & P £ | Total £ |
|---|---|---|---|---|---|---|---|
| Lafarge Double Roman concrete tiles size 418 × 330mm on felt and battens | 500 | m2 | 295 | 4,130 | 5,230 | 1,404 | 10,764 |
| Extra for | | | | | | | |
| cloaked verge course | 60 | m | 24 | 336 | 468 | 121 | 925 |
| eaves vent system | 76 | m | 30 | 420 | 750 | 176 | 1,346 |
| universal ridge tile | 25 | m | 10 | 140 | 100 | 36 | 276 |
| gas vent terminal | 1 | nr | 1 | 14 | 57 | 11 | 82 |
| lead apron | 6 | m | 6 | 84 | 44 | 19 | 147 |
| lead flashing | 12 | m | 10 | 140 | 70 | 32 | 242 |
| lead sill | 6 | m | 6 | 84 | 44 | 19 | 147 |
| | | | **Carried to summary** | | | | 13,928 |

| | Qty | Unit | Labour | Hours £ | Mat'ls £ | O & P £ | Total £ |
|---|---|---|---|---|---|---|---|
| **Lafarge Regent tiles** | | | | | | | |
| **Two-bedroom bungalow** | | | | | | | |
| Lafarge Regent concrete tiles size 418 × 332mm on felt and battens | 180 | m2 | 106 | 1,484 | 1,936 | 513 | 3,933 |
| Extra for | | | | | | | |
| cloaked verge course | 20 | m | 8 | 112 | 156 | 40 | 308 |
| eaves vent system | 36 | m | 14 | 196 | 355 | 83 | 634 |
| universal ridge tile | 18 | m | 7 | 98 | 180 | 42 | 320 |
| gas flue tile | 1 | nr | 1 | 14 | 52 | 10 | 76 |
| | | **Carried to summary** | | | | | 4,962 |
| **Three-bedroom house with one dormer window** | | | | | | | |
| Lafarge Regent concrete tiles size 418 × 332mm on felt and battens | 350 | m2 | 206 | 2,884 | 3,762 | 997 | 7,643 |
| Extra for | | | | | | | |
| eaves vent system | 76 | m | 30 | 420 | 750 | 176 | 1,346 |
| universal ridge tile | 8 | m | 5 | 70 | 104 | 26 | 200 |
| 1/3 hip tile | 48 | m | 19 | 266 | 479 | 112 | 857 |
| | | | | **Carried forward** | | | 10,045 |

| | Qty | Unit | Labour | Hours £ | Mat'ls £ | O & P £ | Total £ |
|---|---|---|---|---|---|---|---|
| | | | | **Brought forward** | | | 10,045 |
| gas flue tile | 1 | nr | 1 | 14 | 52 | 10 | 76 |
| lead apron | 3 | m | 3 | 42 | 22 | 10 | 74 |
| lead flashing | 6 | m | 5 | 70 | 35 | 16 | 121 |
| lead sill | 3 | m | 3 | 42 | 22 | 10 | 74 |
| | | | | | | | 10,389 |
| **Five-bedroom house with two dormer windows** | | | | | | | |
| Lafarge Regent concrete tiles size 418 × 332mm on felt and battens | 500 | m2 | 295 | 5,380 | 5,230 | 1,592 | 12,202 |
| Extra for | | | | | | | |
| cloaked verge course | 60 | m | 24 | 336 | 468 | 121 | 925 |
| eaves vent system | 76 | m | 30 | 420 | 750 | 176 | 1,346 |
| universal ridge tile | 25 | m | 10 | 140 | 100 | 36 | 276 |
| gas vent terminal | 1 | nr | 1 | 14 | 57 | 11 | 82 |
| lead apron | 6 | m | 6 | 84 | 44 | 19 | 147 |
| lead flashing | 12 | m | 10 | 140 | 70 | 32 | 242 |
| lead sill | 6 | m | 6 | 84 | 44 | 19 | 147 |
| | | | **Carried to summary** | | | | 15,365 |

| | Qty | Unit | Labour | Hours £ | Mat'ls £ | O & P £ | Total £ |
|---|---|---|---|---|---|---|---|
| **Lafarge Grovebury tiles** | | | | | | | |
| **Two-bedroom bungalow** | | | | | | | |
| Lafarge Grovebury concrete tiles size 418 × 332mm on felt and battens | 180 | m2 | 106 | 1,484 | 1,936 | 513 | 3,933 |
| Extra for | | | | | | | |
| cloaked verge course | 20 | m | 8 | 112 | 156 | 40 | 308 |
| eaves vent system | 36 | m | 14 | 196 | 355 | 83 | 634 |
| universal ridge tile | 18 | m | 7 | 98 | 180 | 42 | 320 |
| gas flue tile | 1 | nr | 1 | 14 | 52 | 10 | 76 |
| | | **Carried to summary** | | | | | 4,962 |
| **Three-bedroom house with one dormer window** | | | | | | | |
| Lafarge Grovebury concrete tiles size 418 × 332mm on felt and battens | 350 | m2 | 206 | 2,884 | 3,762 | 997 | 7,643 |
| Extra for | | | | | | | |
| eaves vent system | 76 | m | 30 | 420 | 750 | 176 | 1,346 |
| universal ridge tile | 8 | m | 5 | 70 | 104 | 26 | 200 |
| 1/3 hip tile | 48 | m | 19 | 266 | 479 | 112 | 857 |
| | | | | **Carried forward** | | | 10,045 |

| | Qty | Unit | Labour | Hours £ | Mat'ls £ | O & P £ | Total £ |
|---|---|---|---|---|---|---|---|
| | | | | **Brought forward** | | | 10,045 |
| gas flue tile | 1 | nr | 1 | 14 | 52 | 10 | 76 |
| lead apron | 3 | m | 3 | 42 | 22 | 10 | 74 |
| lead flashing | 6 | m | 5 | 70 | 35 | 16 | 121 |
| lead sill | 3 | m | 3 | 42 | 22 | 10 | 74 |
| | | **Carried to summary** | | | | | 10,389 |

**Five-bedroom house with two dormer windows**

| | Qty | Unit | Labour | Hours £ | Mat'ls £ | O & P £ | Total £ |
|---|---|---|---|---|---|---|---|
| Lafarge Grovebury concrete tiles size 418 × 332mm on felt and battens | 500 | m2 | 295 | 5,380 | 5,230 | 1,592 | 12,202 |
| Extra for | | | | | | | |
| cloaked verge course | 60 | m | 24 | 336 | 468 | 121 | 925 |
| eaves vent system | 76 | m | 30 | 420 | 750 | 176 | 1,346 |
| universal ridge tile | 25 | m | 10 | 140 | 100 | 36 | 276 |
| gas vent terminal | 1 | nr | 1 | 14 | 57 | 11 | 82 |
| lead apron | 6 | m | 6 | 84 | 44 | 19 | 147 |
| lead flashing | 12 | m | 10 | 140 | 70 | 32 | 242 |
| lead sill | 6 | m | 6 | 84 | 44 | 19 | 147 |
| | | **Carried to summary** | | | | | 15,365 |

| | Qty | Unit | Labour | Hours £ | Mat'ls £ | O & P £ | Total £ |
|---|---|---|---|---|---|---|---|
| **Lafarge Stonewold tiles** | | | | | | | |
| **Two-bedroom bungalow** | | | | | | | |
| Lafarge Stonewold concrete tiles size 430 × 380mm on felt and battens | 180 | m2 | 106 | 1,484 | 2,858 | 651 | 4,993 |
| Extra for | | | | | | | |
| cloaked verge course | 20 | m | 8 | 112 | 156 | 40 | 308 |
| eaves vent system | 36 | m | 14 | 196 | 355 | 83 | 634 |
| universal ridge tile | 18 | m | 7 | 98 | 180 | 42 | 320 |
| gas flue tile | 1 | nr | 1 | 14 | 52 | 10 | 76 |
| | | **Carried to summary** | | | | | 6,023 |
| **Three-bedroom house with one dormer window** | | | | | | | |
| Lafarge Stonewold concrete tiles size 430 × 380mm on felt and battens | 350 | m2 | 206 | 2,884 | 5,558 | 1,266 | 9,708 |
| Extra for | | | | | | | |
| eaves vent system | 76 | m | 30 | 420 | 750 | 176 | 1,346 |
| universal ridge tile | 8 | m | 5 | 70 | 104 | 26 | 200 |
| 1/3 hip tile | 48 | m | 19 | 266 | 479 | 112 | 857 |
| | | | | **Carried forward** | | | 12,111 |

| | Qty | Unit | Labour | Hours £ | Mat'ls £ | O & P £ | Total £ |
|---|---|---|---|---|---|---|---|
| | | | | **Brought forward** | | | 12,111 |
| gas flue tile | 1 | nr | 1 | 14 | 52 | 10 | 76 |
| lead apron | 3 | m | 3 | 42 | 22 | 10 | 74 |
| lead flashing | 6 | m | 5 | 70 | 35 | 16 | 121 |
| lead sill | 3 | m | 3 | 42 | 22 | 10 | 74 |
| | | **Carried to summary** | | | | | 12,455 |

**Five-bedroom house with two dormer windows**

| | Qty | Unit | Labour | Hours £ | Mat'ls £ | O & P £ | Total £ |
|---|---|---|---|---|---|---|---|
| Lafarge Stonewold concrete tiles size 430 × 380mm on felt and battens | 500 | m2 | 295 | 5,380 | 7,940 | 1,998 | 15,318 |
| Extra for | | | | | | | |
| cloaked verge course | 60 | m | 24 | 336 | 468 | 121 | 925 |
| eaves vent system | 76 | m | 30 | 420 | 750 | 176 | 1,346 |
| universal ridge tile | 25 | m | 10 | 140 | 100 | 36 | 276 |
| gas vent terminal | 1 | nr | 1 | 14 | 57 | 11 | 82 |
| lead apron | 6 | m | 6 | 84 | 44 | 19 | 147 |
| lead flashing | 12 | m | 10 | 140 | 70 | 32 | 242 |
| lead sill | 6 | m | 6 | 84 | 44 | 19 | 147 |
| | | **Carried to summary** | | | | | 18,482 |

| | Qty | Unit | Labour | Hours £ | Mat'ls £ | O & P £ | Total £ |
|---|---|---|---|---|---|---|---|
| **FIBRE CEMENT SLATING** | | | | | | | |
| **Two-bedroom bungalow** | | | | | | | |
| Asbestos-free fibre cement slating 400 × 200mm on felt and battens | 180 | m2 | 162 | 2,268 | 6,427 | 1,304 | 9,999 |
| Extra for | | | | | | | |
| undercloak at verge | 20 | m | 4 | 56 | 42 | 15 | 113 |
| double eaves course | 36 | m | 13 | 182 | 238 | 63 | 483 |
| half round ridge tile | 18 | m | 7 | 98 | 428 | 79 | 605 |
| | | **Carried to summary** | | | | | 11,200 |
| **Three-bedroom house with one dormer window** | | | | | | | |
| Asbestos-free fibre cement slating 400 × 200mm on felt and battens | 350 | m2 | 315 | 4,410 | 12,498 | 2,536 | 19,444 |
| Extra for | | | | | | | |
| double eaves course | 76 | m | 26 | 364 | 501 | 130 | 995 |
| half round ridge tile | 8 | m | 3 | 42 | 190 | 35 | 267 |
| hip tile | 48 | m | 19 | 266 | 1,141 | 211 | 1,618 |
| | | | | **Carried forward** | | | 22,324 |

| | Qty | Unit | Labour | Hours £ | Mat'ls £ | O & P £ | Total £ |
|---|---|---|---|---|---|---|---|
| | | | | **Brought forward** | | | 22,324 |
| lead apron | 3 | m | 3 | 42 | 22 | 10 | 74 |
| lead flashing | 6 | m | 5 | 70 | 35 | 16 | 121 |
| lead sill | 3 | m | 3 | 42 | 22 | 10 | 74 |
| | | | **Carried to summary** | | | | 22,592 |

**Five-bedroom house with two dormer windows**

| | Qty | Unit | Labour | Hours £ | Mat'ls £ | O & P £ | Total £ |
|---|---|---|---|---|---|---|---|
| Asbestos-free fibre cement slating 400 × 200mm on felt and battens | 500 | m2 | 450 | 5,380 | 17,855 | 3,485 | 26,720 |
| Extra for | | | | | | | |
| undercloak at verge | 60 | m | 12 | 168 | 126 | 44 | 338 |
| double eaves course | 76 | m | 26 | 364 | 502 | 130 | 996 |
| half round ridge tile | 25 | m | 10 | 140 | 595 | 110 | 845 |
| lead apron | 6 | m | 6 | 84 | 44 | 19 | 147 |
| lead flashing | 12 | m | 10 | 140 | 70 | 32 | 242 |
| lead sill | 6 | m | 6 | 84 | 44 | 19 | 147 |
| | | | **Carried to summary** | | | | 29,435 |

| | Qty | Unit | Labour | Hours £ | Mat'ls £ | O & P £ | Total £ |
|---|---|---|---|---|---|---|---|
| **Fibre cement slating** | | | | | | | |
| **Two-bedroom bungalow** | | | | | | | |
| Asbestos-free fibre cement slating 500 × 250mm on felt and battens | 180 | m2 | 144 | 2,016 | 5,785 | 1,170 | 8,971 |
| Extra for | | | | | | | |
| undercloak at verge | 20 | m | 4 | 56 | 42 | 15 | 113 |
| double eaves course | 36 | m | 13 | 182 | 238 | 63 | 483 |
| half round ridge tile | 18 | m | 7 | 98 | 428 | 79 | 605 |
| | | **Carried to summary** | | | | | 10,172 |
| **Three-bedroom house with one dormer window** | | | | | | | |
| Asbestos-free fibre cement slating 500 × 250mm on felt and battens | 350 | m2 | 280 | 3,920 | 11,249 | 2,275 | 17,444 |
| Extra for | | | | | | | |
| double eaves course | 76 | m | 26 | 364 | 501 | 130 | 995 |
| half round ridge tile | 8 | m | 3 | 42 | 190 | 35 | 267 |
| hip tile | 48 | m | 19 | 266 | 1,141 | 211 | 1,618 |
| | | | | **Carried forward** | | | 20,324 |

| | Qty | Unit | Labour | Hours £ | Mat'ls £ | O & P £ | Total £ |
|---|---|---|---|---|---|---|---|
| | | | | **Brought forward** | | | 20,324 |
| lead apron | 3 | m | 3 | 42 | 22 | 10 | 74 |
| lead flashing | 6 | m | 5 | 70 | 35 | 16 | 121 |
| lead sill | 3 | m | 3 | 42 | 22 | 10 | 74 |
| | | | **Carried to summary** | | | | 20,592 |

**Five-bedroom house with two dormer windows**

| | Qty | Unit | Labour | Hours £ | Mat'ls £ | O & P £ | Total £ |
|---|---|---|---|---|---|---|---|
| Asbestos-free fibre cement slating 500 × 250mm on felt and battens | 500 | m2 | 400 | 5,380 | 16,070 | 3,218 | 24,668 |
| Extra for | | | | | | | |
| undercloak at verge | 60 | m | 12 | 168 | 126 | 44 | 338 |
| double eaves course | 76 | m | 26 | 364 | 502 | 130 | 996 |
| half round ridge tile | 25 | m | 10 | 140 | 595 | 110 | 845 |
| lead apron | 6 | m | 6 | 84 | 44 | 19 | 147 |
| lead flashing | 12 | m | 10 | 140 | 70 | 32 | 242 |
| lead sill | 6 | m | 6 | 84 | 44 | 19 | 147 |
| | | | **Carried to summary** | | | | 27,383 |

| | Qty | Unit | Labour | Hours £ | Mat'ls £ | O & P £ | Total £ |
|---|---|---|---|---|---|---|---|
| **Fibre cement slating** | | | | | | | |
| **Two-bedroom bungalow** | | | | | | | |
| Asbestos-free fibre cement slating 600 × 300mm on felt and battens | 180 | m2 | 126 | 1,764 | 5,124 | 1,033 | 7,921 |
| Extra for | | | | | | | |
| undercloak at verge | 20 | m | 4 | 56 | 42 | 15 | 113 |
| double eaves course | 36 | m | 13 | 182 | 238 | 63 | 483 |
| half round ridge tile | 18 | m | 7 | 98 | 428 | 79 | 605 |
| | | **Carried to summary** | | | | | 9,122 |
| **Three-bedroom house with one dormer window** | | | | | | | |
| Asbestos-free fibre cement slating 600 × 300mm on felt and battens | 350 | m2 | 245 | 3,430 | 9,965 | 2,009 | 15,404 |
| Extra for | | | | | | | |
| double eaves course | 76 | m | 26 | 364 | 501 | 130 | 995 |
| half round ridge tile | 8 | m | 3 | 42 | 190 | 35 | 267 |
| hip tile | 48 | m | 19 | 266 | 1,141 | 211 | 1,618 |
| | | | | **Carried forward** | | | 18,284 |

| | Qty | Unit | Labour | Hours £ | Mat'ls £ | O & P £ | Total £ |
|---|---|---|---|---|---|---|---|
| | | | | **Brought forward** | | | 18,284 |
| lead apron | 3 | m | 3 | 42 | 22 | 10 | 74 |
| lead flashing | 6 | m | 5 | 70 | 35 | 16 | 121 |
| lead sill | 3 | m | 3 | 42 | 22 | 10 | 74 |
| | | | **Carried to summary** | | | | 18,552 |

**Five-bedroom house with two dormer windows**

| | Qty | Unit | Labour | Hours £ | Mat'ls £ | O & P £ | Total £ |
|---|---|---|---|---|---|---|---|
| Asbestos-free fibre cement slating 600 × 300mm on felt and battens | 500 | m2 | 350 | 5,380 | 14,235 | 2,942 | 22,557 |
| Extra for | | | | | | | |
| undercloak at verge | 60 | m | 12 | 168 | 126 | 44 | 338 |
| double eaves course | 76 | m | 26 | 364 | 502 | 130 | 996 |
| half round ridge tile | 25 | m | 10 | 140 | 595 | 110 | 845 |
| lead apron | 6 | m | 6 | 84 | 44 | 19 | 147 |
| lead flashing | 12 | m | 10 | 140 | 70 | 32 | 242 |
| lead sill | 6 | m | 6 | 84 | 44 | 19 | 147 |
| | | | **Carried to summary** | | | | 25,272 |

| | Qty | Unit | Labour | Hours £ | Mat'ls £ | O & P £ | Total £ |
|---|---|---|---|---|---|---|---|
| **Natural slating** | | | | | | | |
| **Two-bedroom bungalow** | | | | | | | |
| Blue/grey Welsh slating 405 × 205 300mm on felt and battens | 180 | m2 | 216 | 3,024 | 9,518 | 1,881 | 14,423 |
| Extra for | | | | | | | |
| undercloak at verge | 20 | m | 14 | 196 | 371 | 85 | 652 |
| double eaves course | 36 | m | 18 | 252 | 892 | 172 | 1,316 |
| angle ridge tile | 18 | m | 13 | 182 | 289 | 71 | 542 |
| | | **Carried to summary** | | | | | 16,933 |
| **Three-bedroom house with one dormer window** | | | | | | | |
| Blue/grey Welsh slating 405 × 205 300mm on felt and battens | 350 | m2 | 420 | 5,880 | 18,508 | 3,658 | 28,046 |
| Extra for | | | | | | | |
| double eaves course | 76 | m | 53 | 742 | 1,883 | 394 | 3,019 |
| angle ridge tile | 8 | m | 6 | 84 | 128 | 32 | 244 |
| angle hip tile | 48 | m | 34 | 476 | 772 | 187 | 1,435 |
| | | | | **Carried forward** | | | 32,744 |

| | Qty | Unit | Labour | Hours £ | Mat'ls £ | O & P £ | Total £ |
|---|---|---|---|---|---|---|---|
| | | | | **Brought forward** | | | 32,744 |
| lead apron | 3 | m | 3 | 42 | 22 | 10 | 74 |
| lead flashing | 6 | m | 5 | 70 | 35 | 16 | 121 |
| lead sill | 3 | m | 3 | 42 | 22 | 10 | 74 |
| | | | **Carried to summary** | | | | 33,012 |

**Five-bedroom house with two dormer windows**

| | Qty | Unit | Labour | Hours £ | Mat'ls £ | O & P £ | Total £ |
|---|---|---|---|---|---|---|---|
| Blue/grey Welsh slating 405 × 205 300mm on felt and battens | 500 | m2 | 600 | 5,380 | 26,440 | 4,773 | 36,593 |
| Extra for | | | | | | | |
| undercloak at verge | 60 | m | 42 | 588 | 1,113 | 255 | 1,956 |
| double eaves course | 76 | m | 38 | 532 | 1,883 | 362 | 2,777 |
| angle ridge tile | 25 | m | 18 | 252 | 402 | 98 | 752 |
| lead apron | 6 | m | 6 | 84 | 44 | 19 | 147 |
| lead flashing | 12 | m | 10 | 140 | 70 | 32 | 242 |
| lead sill | 6 | m | 6 | 84 | 44 | 19 | 147 |
| | | | **Carried to summary** | | | | 42,614 |

| | Qty | Unit | Labour | Hours £ | Mat'ls £ | O & P £ | Total £ |
|---|---|---|---|---|---|---|---|
| **Natural slating** | | | | | | | |
| **Two-bedroom bungalow** | | | | | | | |
| Blue/grey Welsh slating 510 × 255 300mm on felt and battens | 180 | m2 | 180 | 2,520 | 9,920 | 1,866 | 14,306 |
| Extra for | | | | | | | |
| undercloak at verge | 20 | m | 14 | 196 | 371 | 85 | 652 |
| double eaves course | 36 | m | 18 | 252 | 892 | 172 | 1,316 |
| angle ridge tile | 18 | m | 13 | 182 | 289 | 71 | 542 |
| | | **Carried to summary** | | | | | 16,815 |
| **Three-bedroom house with one dormer window** | | | | | | | |
| Blue/grey Welsh slating 510 × 255 300mm on felt and battens | 350 | m2 | 350 | 4,900 | 19,288 | 3,628 | 27,816 |
| Extra for | | | | | | | |
| double eaves course | 76 | m | 53 | 742 | 1,883 | 394 | 3,019 |
| angle ridge tile | 8 | m | 6 | 84 | 128 | 32 | 244 |
| angle hip tile | 48 | m | 34 | 476 | 772 | 187 | 1,435 |
| | | | | **Carried forward** | | | 32,514 |

| | Qty | Unit | Labour | Hours £ | Mat'ls £ | O & P £ | Total £ |
|---|---|---|---|---|---|---|---|
| | | | | **Brought forward** | | | 32,514 |
| lead apron | 3 | m | 3 | 42 | 22 | 10 | 74 |
| lead flashing | 6 | m | 5 | 70 | 35 | 16 | 121 |
| lead sill | 3 | m | 3 | 42 | 22 | 10 | 74 |
| | | | | | | | 32,782 |
| **Five-bedroom house with two dormer windows** | | | | | | | |
| Blue/grey Welsh slating 510 × 255 300mm on felt and battens | 500 | m2 | 500 | 5,380 | 27,555 | 4,940 | 37,875 |
| Extra for | | | | | | | |
| undercloak at verge | 60 | m | 42 | 588 | 1,113 | 255 | 1,956 |
| double eaves course | 76 | m | 38 | 532 | 1,883 | 362 | 2,777 |
| angle ridge tile | 25 | m | 18 | 252 | 402 | 98 | 752 |
| lead apron | 6 | m | 6 | 84 | 44 | 19 | 147 |
| lead flashing | 12 | m | 10 | 140 | 70 | 32 | 242 |
| lead sill | 6 | m | 6 | 84 | 44 | 19 | 147 |
| | | | **Carried to summary** | | | | 43,897 |

| | Qty | Unit | Labour | Hours £ | Mat'ls £ | O & P £ | Total £ |
|---|---|---|---|---|---|---|---|
| **Natural slating** | | | | | | | |
| **Two-bedroom bungalow** | | | | | | | |
| Blue/grey Welsh slating 610 × 305 300mm on felt and battens | 180 | m2 | 144 | 2,016 | 10,560 | 1,886 | 14,462 |
| Extra for | | | | | | | |
| undercloak at verge | 20 | m | 14 | 196 | 371 | 85 | 652 |
| double eaves course | 36 | m | 18 | 252 | 892 | 172 | 1,316 |
| angle ridge tile | 18 | m | 13 | 182 | 289 | 71 | 542 |
| | | **Carried to summary** | | | | | 16,972 |
| **Three-bedroom house with one dormer window** | | | | | | | |
| Blue/grey Welsh slating 610 × 305 300mm on felt and battens | 280 | m2 | 280 | 3,920 | 20,534 | 3,668 | 28,122 |
| Extra for | | | | | | | |
| double eaves course | 76 | m | 53 | 742 | 1,883 | 394 | 3,019 |
| angle ridge tile | 8 | m | 6 | 84 | 128 | 32 | 244 |
| angle hip tile | 48 | m | 34 | 476 | 772 | 187 | 1,435 |
| | | | | **Carried forward** | | | 32,820 |

| | Qty | Unit | Labour | Hours £ | Mat'ls £ | O & P £ | Total £ |
|---|---|---|---|---|---|---|---|
| | | | | **Brought forward** | | | 32,820 |
| lead apron | 3 | m | 3 | 42 | 22 | 10 | 74 |
| lead flashing | 6 | m | 5 | 70 | 35 | 16 | 121 |
| lead sill | 3 | m | 3 | 42 | 22 | 10 | 74 |
| | | | **Carried to summary** | | | | 33,088 |

**Five-bedroom house with two dormer windows**

| | Qty | Unit | Labour | Hours £ | Mat'ls £ | O & P £ | Total £ |
|---|---|---|---|---|---|---|---|
| Blue/grey Welsh slating 610 × 305 300mm on felt and battens | 500 | m2 | 400 | 5,380 | 29,335 | 5,207 | 39,922 |
| Extra for | | | | | | | |
| undercloak at verge | 60 | m | 42 | 588 | 1,113 | 255 | 1,956 |
| double eaves course | 76 | m | 38 | 532 | 1,883 | 362 | 2,777 |
| angle ridge tile | 25 | m | 18 | 252 | 402 | 98 | 752 |
| lead apron | 6 | m | 6 | 84 | 44 | 19 | 147 |
| lead flashing | 12 | m | 10 | 140 | 70 | 32 | 242 |
| lead sill | 6 | m | 6 | 84 | 44 | 19 | 147 |
| | | | **Carried to summary** | | | | 45,944 |

| | Qty | Unit | Labour | Hours £ | Mat'ls £ | O & P £ | Total £ |
|---|---|---|---|---|---|---|---|
| **Natural slating** | | | | | | | |
| **Two-bedroom bungalow** | | | | | | | |
| Westmorland slating in random lengths 450mm to 330mm on felt and battens | 180 | m2 | 243 | 3,402 | 21,277 | 3,702 | 28,381 |
| Extra for | | | | | | | |
| undercloak at verge | 20 | m | 14 | 196 | 371 | 85 | 652 |
| double eaves course | 36 | m | 18 | 252 | 892 | 172 | 1,316 |
| angle ridge tile | 18 | m | 13 | 182 | 289 | 71 | 542 |
| | | **Carried to summary** | | | | | 30,890 |
| **Three-bedroom house with one dormer window** | | | | | | | |
| Westmorland slating in random lengths 450mm to 330mm on felt and battens | 280 | m2 | 472 | 6,608 | 41,373 | 7,197 | 55,178 |
| Extra for | | | | | | | |
| double eaves course | 76 | m | 53 | 742 | 1,883 | 394 | 3,019 |
| angle ridge tile | 8 | m | 6 | 84 | 128 | 32 | 244 |
| angle hip tile | 48 | m | 34 | 476 | 772 | 187 | 1,435 |
| | | | | **Carried forward** | | | 59,876 |

| | Qty | Unit | Labour | Hours £ | Mat'ls £ | O & P £ | Total £ |
|---|---|---|---|---|---|---|---|
| | | | | **Brought forward** | | | 59,876 |
| lead apron | 3 | m | 3 | 42 | 22 | 10 | 74 |
| lead flashing | 6 | m | 5 | 70 | 35 | 16 | 121 |
| lead sill | 3 | m | 3 | 42 | 22 | 10 | 74 |
| | | | | | | | 60,144 |
| **Five-bedroom house with two dormer windows** | | | | | | | |
| Westmorland slating in random lengths 450mm to 330mm on felt and battens | 500 | m2 | 675 | 5,380 | 59,105 | 9,673 | 74,158 |
| Extra for | | | | | | | |
| undercloak at verge | 60 | m | 42 | 588 | 1,113 | 255 | 1,956 |
| double eaves course | 76 | m | 38 | 532 | 1,883 | 362 | 2,777 |
| angle ridge tile | 25 | m | 18 | 252 | 402 | 98 | 752 |
| lead apron | 6 | m | 6 | 84 | 44 | 19 | 147 |
| lead flashing | 12 | m | 10 | 140 | 70 | 32 | 242 |
| lead sill | 6 | m | 6 | 84 | 44 | 19 | 147 |
| | | | **Carried to summary** | | | | 80,179 |

| | Qty | Unit | Labour | Hours £ | Mat'ls £ | O & P £ | Total £ |
|---|---|---|---|---|---|---|---|
| **Reconstructed stone slating** | | | | | | | |
| **Two-bedroom bungalow** | | | | | | | |
| Marley Monarch interlocking slate 325 × 330mm on felt and battens | 180 | m2 | 108 | 1,512 | 4,813 | 949 | 7,274 |
| Extra for | | | | | | | |
| interlocking dry verge | 20 | m | 12 | 168 | 97 | 40 | 305 |
| eaves vent system | 36 | m | 22 | 308 | 465 | 116 | 889 |
| Modern ridge tile | 18 | m | 7 | 98 | 132 | 35 | 265 |
| | | **Carried to summary** | | | | | 8,732 |
| **Three-bedroom house with one dormer window** | | | | | | | |
| Marley Monarch interlocking slate 325 × 330mm on felt and battens | 280 | m2 | 210 | 2,940 | 9,359 | 1,845 | 14,144 |
| Extra for | | | | | | | |
| eaves vent system | 76 | m | 46 | 644 | 983 | 244 | 1,871 |
| Modern ridge tile | 8 | m | 3 | 42 | 59 | 15 | 116 |
| Modern hip tile | 48 | m | 19 | 266 | 353 | 93 | 712 |
| | | | | **Carried forward** | | | 16,843 |

| | Qty | Unit | Labour | Hours £ | Mat'ls £ | O & P £ | Total £ |
|---|---|---|---|---|---|---|---|
| | | | | **Brought forward** | | | 16,843 |
| lead apron | 3 | m | 3 | 42 | 22 | 10 | 74 |
| lead flashing | 6 | m | 5 | 70 | 35 | 16 | 121 |
| lead sill | 3 | m | 3 | 42 | 22 | 10 | 74 |
| | | | **Carried to summary** | | | | 17,111 |

**Five-bedroom house with two dormer windows**

| | Qty | Unit | Labour | Hours £ | Mat'ls £ | O & P £ | Total £ |
|---|---|---|---|---|---|---|---|
| Marley Monarch interlocking slate 325 × 330mm on felt and battens | 500 | m2 | 300 | 5,380 | 13,370 | 2,813 | 21,563 |
| Extra for | | | | | | | |
| interlocking dry verge | 60 | m | 36 | 504 | 291 | 119 | 914 |
| eaves vent system | 76 | m | 46 | 644 | 983 | 244 | 1,871 |
| Modern ridge tile | 25 | m | 10 | 140 | 184 | 49 | 373 |
| lead apron | 6 | m | 6 | 84 | 44 | 19 | 147 |
| lead flashing | 12 | m | 10 | 140 | 70 | 32 | 242 |
| lead sill | 6 | m | 6 | 84 | 44 | 19 | 147 |
| | | | **Carried to summary** | | | | 25,256 |

| | Qty | Unit | Labour | Hours £ | Mat'ls £ | O & P £ | Total £ |
|---|---|---|---|---|---|---|---|
| **Reconstructed stone slating** | | | | | | | |
| **Two-bedroom bungalow** | | | | | | | |
| Lafarge Cambrian coloured slating 300 × 336mm on felt and battens | 180 | m2 | 108 | 1,512 | 4,554 | 910 | 6,976 |
| Extra for | | | | | | | |
| verge system | 20 | m | 8 | 112 | 220 | 50 | 382 |
| eaves vent system | 36 | m | 22 | 308 | 349 | 99 | 756 |
| half-round ridge tile | 18 | m | 7 | 98 | 180 | 42 | 320 |
| **Carried to summary** | | | | | | | 8,433 |
| **Three-bedroom house with one dormer window** | | | | | | | |
| Lafarge Cambrian coloured slating 300 × 336mm on felt and battens | 280 | m2 | 210 | 2,940 | 8,855 | 1,769 | 13,564 |
| Extra for | | | | | | | |
| eaves vent system | 76 | m | 46 | 644 | 737 | 207 | 1,588 |
| half-round ridge tile | 8 | m | 1 | 14 | 80 | 14 | 108 |
| half-round hip tile | 48 | m | 19 | 266 | 480 | 112 | 858 |
| **Carried forward** | | | | | | | 16,118 |

| | Qty | Unit | Labour | Hours £ | Mat'ls £ | O & P £ | Total £ |
|---|---|---|---|---|---|---|---|
| | | | | **Brought forward** | | | 16,118 |
| lead apron | 3 | m | 3 | 42 | 22 | 10 | 74 |
| lead flashing | 6 | m | 5 | 70 | 35 | 16 | 121 |
| lead sill | 3 | m | 3 | 42 | 22 | 10 | 74 |
| | | | **Carried to summary** | | | | 16,386 |

**Five-bedroom house with two dormer windows**

| | Qty | Unit | Labour | Hours £ | Mat'ls £ | O & P £ | Total £ |
|---|---|---|---|---|---|---|---|
| Lafarge Cambrian coloured slating 300 × 336mm on felt and battens | 500 | m2 | 300 | 5,380 | 12,650 | 2,705 | 20,735 |
| Extra for | | | | | | | |
| verge system | 60 | m | 24 | 336 | 661 | 150 | 1,147 |
| eaves vent system | 76 | m | 46 | 644 | 737 | 207 | 1,588 |
| half-round ridge tile | 25 | m | 10 | 140 | 250 | 59 | 449 |
| lead apron | 6 | m | 6 | 84 | 44 | 19 | 147 |
| lead flashing | 12 | m | 10 | 140 | 70 | 32 | 242 |
| lead sill | 6 | m | 6 | 84 | 44 | 19 | 147 |
| | | | **Carried to summary** | | | | 24,454 |

# SUMMARY OF PROJECT COSTS

## CLAY/CONCRETE ROOF TILING

| | Total £ |
|---|---|
| **Marley Marlden tiles** | |
| Two-bedroom bungalow | 12,657 |
| Three-bedroom house with one dormer window | 24,492 |
| Five-bedroom house with two dormer windows | 34,944 |
| **Marley Thaxton tiles** | |
| Two-bedroom bungalow | 12,538 |
| Three-bedroom house with one dormer window | 24,744 |
| Five-bedroom house with two dormer windows | 35,220 |
| **Marley Heritage tiles** | |
| Two-bedroom bungalow | 12,593 |
| Three-bedroom house with one dormer window | 24,591 |
| Five-bedroom house with two dormer windows | 35,369 |
| **Marley Plain tiles** | |
| Two-bedroom bungalow | 12,365 |
| Three-bedroom house with one dormer window | 24,149 |
| Five-bedroom house with two dormer windows | 34,737 |
| **Marley Modern tiles** | |
| Two-bedroom bungalow | 5,454 |
| Three-bedroom house with one dormer window | 10,633 |
| Five-bedroom house with two dormer windows | 13,434 |

| | Total £ |
|---|---|
| **Marley Ludlow Major tiles** | |
| Two-bedroom bungalow | 5,473 |
| Three-bedroom house with one dormer window | 10,825 |
| Five-bedroom house with two dormer windows | 13,485 |
| **Marley Ludlow Plus tiles** | |
| Two-bedroom bungalow | 6,210 |
| Three-bedroom house with one dormer window | 11,800 |
| Five-bedroom house with two dormer windows | 15,722 |
| **Marley Anglia Plus tiles** | |
| Two-bedroom bungalow | 6,668 |
| Three-bedroom house with one dormer window | 12,690 |
| Five-bedroom house with two dormer windows | 16,992 |
| **Marley Mendip tiles** | |
| Two-bedroom bungalow | 5,710 |
| Three-bedroom house with one dormer window | 10,619 |
| Five-bedroom house with two dormer windows | 14,330 |
| **Marley Malvern tiles** | |
| Two-bedroom bungalow | 5,629 |
| Three-bedroom house with one dormer window | 10,669 |
| Five-bedroom house with two dormer windows | 14,106 |
| **Marley Double Roman tiles** | |
| Two-bedroom bungalow | 5,720 |
| Three-bedroom house with one dormer window | 10,846 |
| Five-bedroom house with two dormer windows | 14,359 |

| | **Total** **£** |
|---|---|
| **Marley Bold Roll tiles** | |
| Two-bedroom bungalow | 5,560 |
| Three-bedroom house with one dormer window | 10,536 |
| Five-bedroom house with two dormer windows | 13,916 |
| **Marley Wessex tiles** | |
| Two-bedroom bungalow | 6,469 |
| Three-bedroom house with one dormer window | 12,303 |
| Five-bedroom house with two dormer windows | 16,440 |
| **Lafarge Norfolk tiles** | |
| Two-bedroom bungalow | 5,339 |
| Three-bedroom house with one dormer window | 11,126 |
| Five-bedroom house with two dormer windows | 15,147 |
| **Lafarge Double Roman tiles** | |
| Two-bedroom bungalow | 4,901 |
| Three-bedroom house with one dormer window | 10,273 |
| Five-bedroom house with two dormer windows | 13,928 |
| **Lafarge Regent tiles** | |
| Two-bedroom bungalow | 4,962 |
| Three-bedroom house with one dormer window | 10,389 |
| Five-bedroom house with two dormer windows | 15,365 |
| **Lafarge Grovebury tiles** | |
| Two-bedroom bungalow | 4,962 |
| Three-bedroom house with one dormer window | 10,389 |
| Five-bedroom house with two dormer windows | 15,365 |

| | Total £ |
|---|---|
| **Lafarge Stonewold tiles** | |
| Two-bedroom bungalow | 6,023 |
| Three-bedroom house with one dormer window | 12,455 |
| Five-bedroom house with two dormer windows | 18,482 |
| **FIBRE CEMENT SLATING** | |
| **Asbestos-free 400 × 200mm slates** | |
| Two-bedroom bungalow | 11,200 |
| Three-bedroom house with one dormer window | 22,592 |
| Five-bedroom house with two dormer windows | 29,435 |
| **Asbestos-free 500 × 250mm slates** | |
| Two-bedroom bungalow | 10,172 |
| Three-bedroom house with one dormer window | 20,592 |
| Five-bedroom house with two dormer windows | 27,383 |
| **Asbestos-free 600 × 300mm slates** | |
| Two-bedroom bungalow | 9,122 |
| Three-bedroom house with one dormer window | 18,552 |
| Five-bedroom house with two dormer windows | 25,272 |
| **NATURAL SLATING** | |
| **Blue/grey Welsh 405 × 205mm slates** | |
| Two-bedroom bungalow | 16,933 |
| Three-bedroom house with one dormer window | 33,012 |
| Five-bedroom house with two dormer windows | 42,614 |

| | **Total** £ |
|---|---|
| **Blue/grey Welsh 510 × 255mm slates** | |
| Two-bedroom bungalow | 16,815 |
| Three-bedroom house with one dormer window | 32,782 |
| Five-bedroom house with two dormer windows | 43,897 |
| **Blue/grey Welsh 610 × 305mm slates** | |
| Two-bedroom bungalow | 16,972 |
| Three-bedroom house with one dormer window | 33,088 |
| Five-bedroom house with two dormer windows | 45,944 |
| **Westmorland green random lengths slates** | |
| Two-bedroom bungalow | 30,890 |
| Three-bedroom house with one dormer window | 60,144 |
| Five-bedroom house with two dormer windows | 80,179 |
| **RECONSTRUCTED STONE SLATING** | |
| **Marley Monarch slates** | |
| Two-bedroom bungalow | 8,732 |
| Three-bedroom house with one dormer window | 17,111 |
| Five-bedroom house with two dormer windows | 25,256 |
| **Lafarge Cambrian slates** | |
| Two-bedroom bungalow | 8,433 |
| Three-bedroom house with one dormer window | 16,386 |
| Five-bedroom house with two dormer windows | 24,454 |

# Part Three

## ALTERATIONS AND REPAIRS

| | Unit | Labour | Hours £ | Mat'ls £ | O & P £ | Total £ |
|---|---|---|---|---|---|---|
| **ALTERATIONS AND REPAIRS** | | | | | | |
| Take up roof coverings from pitched roof | | | | | | |
| tiles | m2 | 0.80 | 11.20 | 0.00 | 1.68 | 12.88 |
| slates | m2 | 0.80 | 11.20 | 0.00 | 1.68 | 12.88 |
| timber boarding | m2 | 1.00 | 14.00 | 0.00 | 2.10 | 16.10 |
| metal sheeting | m2 | 0.20 | 2.80 | 0.00 | 0.42 | 3.22 |
| flat sheeting | m2 | 0.30 | 4.20 | 0.00 | 0.63 | 4.83 |
| underfelt | m2 | 0.10 | 1.40 | 0.00 | 0.21 | 1.61 |
| Take up roof coverings from flat roof | | | | | | |
| bituminous felt | m2 | 0.25 | 3.50 | 0.00 | 0.53 | 4.03 |
| metal sheeting | m2 | 0.30 | 4.20 | 0.00 | 0.63 | 4.83 |
| woodwool slabs | m2 | 0.50 | 7.00 | 0.00 | 1.05 | 8.05 |
| firrings | m2 | 0.20 | 2.80 | 0.00 | 0.42 | 3.22 |
| Take up roof coverings from pitched roof, carefully lay aside for reuse | | | | | | |
| tiles | m2 | 0.80 | 11.20 | 0.00 | 1.68 | 12.88 |
| slates | m2 | 0.80 | 11.20 | 0.00 | 1.68 | 12.88 |
| metal sheeting | m2 | 0.50 | 7.00 | 0.00 | 1.05 | 8.05 |
| flat sheeting | m2 | 0.60 | 8.40 | 0.00 | 1.26 | 9.66 |
| Take up roof coverings from flat roof, carefully lay aside for reuse | | | | | | |
| metal sheeting | m2 | 0.60 | 8.40 | 0.00 | 1.26 | 9.66 |
| woodwool slabs | m2 | 0.80 | 11.20 | 0.00 | 1.68 | 12.88 |

| | Unit | Labour Hours | £ | Mat'ls £ | O & P £ | Total £ |
|---|---|---|---|---|---|---|
| **Fibre cement cladding** | | | | | | |
| Take down damaged corrugated fibre cement sheet cladding in sloping areas less than 5m2 and replace with new | | | | | | |
| profile 3, grey | | | | | | |
| fixed to timber purlins | m2 | 0.48 | 6.72 | 14.82 | 3.23 | 24.77 |
| fixed to steel purlins | m2 | 0.54 | 7.56 | 15.11 | 3.40 | 26.07 |
| profile 3, coloured | | | | | | |
| fixed to timber purlins | m2 | 0.48 | 6.72 | 16.56 | 3.49 | 26.77 |
| fixed to steel purlins | m2 | 0.54 | 7.56 | 16.89 | 3.67 | 28.12 |
| profile 6, grey | | | | | | |
| fixed to timber purlins | m2 | 0.54 | 7.56 | 15.32 | 3.43 | 26.31 |
| fixed to steel purlins | m2 | 0.60 | 8.40 | 15.65 | 3.61 | 27.66 |
| profile 6, coloured | | | | | | |
| fixed to timber purlins | m2 | 0.54 | 7.56 | 15.80 | 3.50 | 26.86 |
| fixed to steel purlins | m2 | 0.60 | 8.40 | 16.23 | 3.69 | 28.32 |
| Take down damaged corrugated fibre cement sheet cladding in vertical areas less than 5m2 and replace with new | | | | | | |
| profile 3, grey | | | | | | |
| fixed to timber purlins | m2 | 0.57 | 7.98 | 14.82 | 3.42 | 26.22 |
| fixed to steel purlins | m2 | 0.63 | 8.82 | 15.11 | 3.59 | 27.52 |

| | Unit | Labour | Hours £ | Mat'ls £ | O & P £ | Total £ |
|---|---|---|---|---|---|---|
| **Take down damaged corrugated fibre cement cladding (cont'd)** | | | | | | |
| profile 3, coloured | | | | | | |
| fixed to timber purlins | m2 | 0.57 | 7.98 | 16.56 | 3.68 | 28.22 |
| fixed to steel purlins | m2 | 0.63 | 8.82 | 16.89 | 3.86 | 29.57 |
| profile 6, grey | | | | | | |
| fixed to timber purlins | m2 | 0.63 | 8.82 | 15.32 | 3.62 | 27.76 |
| fixed to steel purlins | m2 | 0.69 | 9.66 | 15.65 | 3.80 | 29.11 |
| profile 6, coloured | | | | | | |
| fixed to timber purlins | m2 | 0.63 | 8.82 | 15.80 | 3.69 | 28.31 |
| fixed to steel purlins | m2 | 0.69 | 9.66 | 16.23 | 3.88 | 29.77 |
| **Galvanised steel sheeting** | | | | | | |
| Take down damaged galvanised steel strip troughed sheeting with standard light grey internal coat and Plastisol external coat in sloping areas less than 5m2 and replace with new | | | | | | |
| 7mm thick with 19mm corrugations | m2 | 0.54 | 7.56 | 10.74 | 2.75 | 21.05 |
| 7mm thick with 32mm corrugations | m2 | 0.60 | 8.40 | 10.88 | 2.89 | 22.17 |
| 7mm thick with 38mm corrugations | m2 | 0.66 | 9.24 | 11.00 | 3.04 | 23.28 |
| 7mm thick with 46mm corrugations | m2 | 0.72 | 10.08 | 11.20 | 3.19 | 24.47 |
| 7mm thick with 60mm corrugations | m2 | 0.78 | 10.92 | 11.45 | 3.36 | 25.73 |

| | Unit | Labour Hours | £ | Mat'ls £ | O & P £ | Total £ |
|---|---|---|---|---|---|---|
| Take down damaged galvanised steel strip troughed sheeting with standard light grey internal coat and Plastisol external coat in vertical areas less than 5m2 and replace with new | | | | | | |
| 7mm thick with 19mm corrugations | m2 | 0.60 | 8.40 | 10.74 | 2.87 | 22.01 |
| 7mm thick with 32mm corrugations | m2 | 0.66 | 9.24 | 10.88 | 3.02 | 23.14 |
| 7mm thick with 38mm corrugations | m2 | 0.72 | 10.08 | 11.00 | 3.16 | 24.24 |
| 7mm thick with 46mm corrugations | m2 | 0.78 | 10.92 | 11.20 | 3.32 | 25.44 |
| 7mm thick with 60mm corrugations | m2 | 0.84 | 11.76 | 11.45 | 3.48 | 26.69 |

**Translucent sheeting**

| | Unit | Labour Hours | £ | Mat'ls £ | O & P £ | Total £ |
|---|---|---|---|---|---|---|
| Take down damaged glass fibre reinforced translucent sheeting with with one corrugation side lap and 150mm top and bottom laps in sloping areas less than 5m2 and replace with new | | | | | | |
| profile 3 sheets | | | | | | |
| to timber members | m2 | 0.48 | 6.72 | 15.24 | 3.29 | 25.25 |
| to steel members | m2 | 0.54 | 7.56 | 15.45 | 3.45 | 26.46 |

| | Unit | Labour | Hours £ | Mat'ls £ | O & P £ | Total £ |
|---|---|---|---|---|---|---|
| **Take down damaged translucent sheeting (cont'd)** | | | | | | |
| profile 6 sheets | | | | | | |
| to timber members | m2 | 0.54 | 7.56 | 15.24 | 3.42 | 26.22 |
| to steel members | m2 | 0.60 | 8.40 | 15.45 | 3.58 | 27.43 |
| Take down damaged glass fibre reinforced translucent sheeting with with one corrugation side lap and 150mm top and bottom laps in vertical areas less than 5m2 and replace with new | | | | | | |
| profile 3 sheets | | | | | | |
| to timber members | m2 | 0.54 | 7.56 | 15.24 | 3.42 | 26.22 |
| to steel members | m2 | 0.60 | 8.40 | 15.45 | 3.58 | 27.43 |
| profile 6 sheets | | | | | | |
| to timber members | m2 | 0.60 | 8.40 | 15.24 | 3.55 | 27.19 |
| to steel members | m2 | 0.66 | 9.24 | 15.45 | 3.70 | 28.39 |
| **Roof battens** | | | | | | |
| Inspect roof battens, refix loose and replace with new, size 38 × 19mm | | | | | | |
| 25% of area | | | | | | |
| 250mm centres | m2 | 0.14 | 1.96 | 0.26 | 0.33 | 2.55 |
| 450mm centres | m2 | 0.12 | 1.68 | 0.19 | 0.28 | 2.15 |
| 600mm centres | m2 | 0.10 | 1.40 | 0.13 | 0.23 | 1.76 |

| | Unit | Labour Hours | Hours £ | Mat'ls £ | O & P £ | Total £ |
|---|---|---|---|---|---|---|
| **Inspect roof battens, refix loose (cont'd)** | | | | | | |
| 50% of area | | | | | | |
| 250mm centres | m2 | 0.26 | 3.64 | 0.52 | 0.62 | 4.78 |
| 450mm centres | m2 | 0.16 | 2.24 | 0.36 | 0.39 | 2.99 |
| 600mm centres | m2 | 0.18 | 2.52 | 0.26 | 0.42 | 3.20 |
| 75% of area | | | | | | |
| 250mm centres | m2 | 0.36 | 5.04 | 0.78 | 0.87 | 6.69 |
| 450mm centres | m2 | 0.26 | 3.64 | 0.55 | 0.63 | 4.82 |
| 600mm centres | m2 | 0.22 | 3.08 | 0.36 | 0.52 | 3.96 |
| 100% of area | | | | | | |
| 250mm centres | m2 | 0.44 | 6.16 | 1.04 | 1.08 | 8.28 |
| 450mm centres | m2 | 0.32 | 4.48 | 0.78 | 0.79 | 6.05 |
| 600mm centres | m2 | 0.38 | 5.32 | 0.52 | 0.88 | 6.72 |
| Inspect roof battens, refix loose and replace with new, size 38 × 25mm | | | | | | |
| 25% of area | | | | | | |
| 250mm centres | m2 | 0.14 | 1.96 | 0.36 | 0.35 | 2.67 |
| 450mm centres | m2 | 0.12 | 1.68 | 0.27 | 0.29 | 2.24 |
| 600mm centres | m2 | 0.10 | 1.40 | 0.18 | 0.24 | 1.82 |
| 50% of area | | | | | | |
| 250mm centres | m2 | 0.26 | 3.64 | 0.72 | 0.65 | 5.01 |
| 450mm centres | m2 | 0.16 | 2.24 | 0.54 | 0.42 | 3.20 |
| 600mm centres | m2 | 0.18 | 2.52 | 0.36 | 0.43 | 3.31 |
| 75% of area | | | | | | |
| 250mm centres | m2 | 0.36 | 5.04 | 1.08 | 0.92 | 7.04 |
| 450mm centres | m2 | 0.26 | 3.64 | 0.81 | 0.67 | 5.12 |
| 600mm centres | m2 | 0.22 | 3.08 | 0.54 | 0.54 | 4.16 |

| | Unit | Labour Hours | £ | Mat'ls £ | O & P £ | Total £ |
|---|---|---|---|---|---|---|
| 100% of area | | | | | | |
| 250mm centres | m2 | 0.44 | 6.16 | 1.44 | 1.14 | 8.74 |
| 450mm centres | m2 | 0.32 | 4.48 | 1.08 | 0.83 | 6.39 |
| 600mm centres | m2 | 0.38 | 5.32 | 0.72 | 0.91 | 6.95 |
| Inspect roof battens, refix loose and replace with new, size 38 × 38mm | | | | | | |
| 25% of area | | | | | | |
| 250mm centres | m2 | 0.14 | 1.96 | 0.54 | 0.38 | 2.88 |
| 450mm centres | m2 | 0.12 | 1.68 | 0.40 | 0.31 | 2.39 |
| 600mm centres | m2 | 0.10 | 1.40 | 0.27 | 0.25 | 1.92 |
| 50% of area | | | | | | |
| 250mm centres | m2 | 0.26 | 3.64 | 1.08 | 0.71 | 5.43 |
| 450mm centres | m2 | 0.16 | 2.24 | 0.81 | 0.46 | 3.51 |
| 600mm centres | m2 | 0.18 | 2.52 | 0.54 | 0.46 | 3.52 |
| 75% of area | | | | | | |
| 250mm centres | m2 | 0.36 | 5.04 | 1.62 | 1.00 | 7.66 |
| 450mm centres | m2 | 0.26 | 3.64 | 1.21 | 0.73 | 5.58 |
| 600mm centres | m2 | 0.22 | 3.08 | 0.81 | 0.58 | 4.47 |
| 100% of area | | | | | | |
| 250mm centres | m2 | 0.44 | 6.16 | 2.16 | 1.25 | 9.57 |
| 450mm centres | m2 | 0.32 | 4.48 | 1.62 | 0.92 | 7.02 |
| 600mm centres | m2 | 0.38 | 5.32 | 1.08 | 0.96 | 7.36 |

| | Unit | Labour | Hours £ | Mat'ls £ | O & P £ | Total £ |
|---|---|---|---|---|---|---|
| **Inspect roof battens, refix loose (cont'd)** | | | | | | |
| Inspect roof battens, refix loose and replace with new, size 44 × 38mm | | | | | | |
| 25% of area | | | | | | |
| 250mm centres | m2 | 0.14 | 1.96 | 0.62 | 0.39 | 2.97 |
| 450mm centres | m2 | 0.12 | 1.68 | 0.47 | 0.32 | 2.47 |
| 600mm centres | m2 | 0.10 | 1.40 | 0.31 | 0.26 | 1.97 |
| 50% of area | | | | | | |
| 250mm centres | m2 | 0.26 | 3.64 | 1.24 | 0.73 | 5.61 |
| 450mm centres | m2 | 0.16 | 2.24 | 0.93 | 0.48 | 3.65 |
| 600mm centres | m2 | 0.18 | 2.52 | 0.62 | 0.47 | 3.61 |
| 75% of area | | | | | | |
| 250mm centres | m2 | 0.36 | 5.04 | 1.86 | 1.04 | 7.94 |
| 450mm centres | m2 | 0.26 | 3.64 | 1.40 | 0.76 | 5.80 |
| 600mm centres | m2 | 0.22 | 3.08 | 0.93 | 0.60 | 4.61 |
| 100% of area | | | | | | |
| 250mm centres | m2 | 0.44 | 6.16 | 2.48 | 1.30 | 9.94 |
| 450mm centres | m2 | 0.32 | 4.48 | 1.86 | 0.95 | 7.29 |
| 600mm centres | m2 | 0.38 | 5.32 | 1.24 | 0.98 | 7.54 |
| **Roof tiling** | | | | | | |
| Remove single broken tile or slate and replace with new | | | | | | |
| Marley Marlden Plain size 267 x 168mm | nr | 0.32 | 4.48 | 0.52 | 0.75 | 5.75 |
| Marley Thaxton Plain size 270 x 168mm | nr | 0.32 | 4.48 | 0.53 | 0.75 | 5.76 |

| | Unit | Labour | Hours | Mat'ls | O & P | Total |
|---|---|---|---|---|---|---|
| | | | £ | £ | £ | £ |
| Marley Heritage Plain size x 168mm | nr | 0.32 | 4.48 | 0.53 | 0.75 | 5.76 |
| Marley Plain size 267 x 168mm | nr | 0.32 | 4.48 | 0.51 | 0.75 | 5.74 |
| Marley Modern Plain size 420 x 330mm | nr | 0.32 | 4.48 | 0.78 | 0.79 | 6.05 |
| Marley Ludlow Major size 267 x 168mm | nr | 0.32 | 4.48 | 0.79 | 0.79 | 6.06 |
| Marley Ludlow Plus size 387 x 229mm | nr | 0.32 | 4.48 | 0.57 | 0.76 | 5.81 |
| Marley Anglia Plus size 387 x 230mm | nr | 0.32 | 4.48 | 0.71 | 0.78 | 5.97 |
| Marley Modern Plain size 420 x 330mm | nr | 0.32 | 4.48 | 0.82 | 0.80 | 6.10 |
| Marley Modern Plain size 420 x 330mm | nr | 0.32 | 4.48 | 0.77 | 0.79 | 6.04 |
| Marley Modern Plain size 420 x 330mm | nr | 0.32 | 4.48 | 0.78 | 0.79 | 6.05 |
| Marley Modern Plain size 420 x 330mm | nr | 0.32 | 4.48 | 0.80 | 0.79 | 6.07 |
| Marley Modern Plain size 420 x 330mm | nr | 0.32 | 4.48 | 1.30 | 0.87 | 6.65 |
| Lafarge Norfolk size 381 x 227mm | nr | 0.32 | 4.48 | 0.60 | 0.76 | 5.84 |
| Lafarge Renown size 418 x 330mm | nr | 0.32 | 4.48 | 0.79 | 0.79 | 6.06 |
| Lafarge Double Roman size 418 x 330mm | nr | 0.32 | 4.48 | 0.79 | 0.79 | 6.06 |
| Lafarge Regent size 418 x 332mm | nr | 0.32 | 4.48 | 0.82 | 0.80 | 6.10 |
| Lafarge Grovebury size 418 x 332mm | nr | 0.32 | 4.48 | 0.82 | 0.80 | 6.10 |
| Lafarge Stonewold size 430 x 380mm | nr | 0.32 | 4.48 | 1.56 | 0.91 | 6.95 |
| Lafarge Cambria size 330 x 336mm | nr | 0.32 | 4.48 | 1.81 | 0.94 | 7.23 |

| | Unit | Labour Hours | £ | Mat'ls £ | O & P £ | Total £ |
|---|---|---|---|---|---|---|
| **Remove single broken tile or slate and replace with new (cont'd)** | | | | | | |
| fibre-cement slate size 400 x 200mm | nr | 0.32 | 4.48 | 1.55 | 0.90 | 6.93 |
| fibre-cement slate size 500 x 250mm | nr | 0.32 | 4.48 | 2.63 | 1.07 | 8.18 |
| fibre-cement slate size 600 x 300mm | nr | 0.32 | 4.48 | 4.53 | 1.35 | 10.36 |
| Welsh Blue slate size 405 x 205mm | nr | 0.32 | 4.48 | 1.83 | 0.95 | 7.26 |
| Welsh Blue slate size 510 x 255mm | nr | 0.32 | 4.48 | 3.06 | 1.13 | 8.67 |
| Welsh Blue slate size 610 x 305mm | nr | 0.32 | 4.48 | 4.89 | 1.41 | 10.78 |
| Westmorland slate size 450 x 230mm | nr | 0.32 | 4.48 | 5.14 | 1.44 | 11.06 |
| Remove existing tiles or slates and replace with new in areas less than 1m2 | | | | | | |
| Marley Marlden Plain size 267 x 168mm | m2 | 2.16 | 30.24 | 33.76 | 9.60 | 73.60 |
| Marley Thaxton Plain size 270 x 168mm | m2 | 2.16 | 30.24 | 34.24 | 9.67 | 74.15 |
| Marley Heritage Plain size x 168mm | m2 | 2.16 | 30.24 | 34.50 | 9.71 | 74.45 |
| Marley Plain size 267 x 168mm | m2 | 2.16 | 30.24 | 33.40 | 9.55 | 73.19 |
| Marley Modern Plain size 420 x 330mm | m2 | 0.81 | 11.34 | 10.47 | 3.27 | 25.08 |
| Marley Ludlow Major size 267 x 168mm | m2 | 0.81 | 11.34 | 10.56 | 3.29 | 25.19 |

| | Unit | Labour Hours | £ | Mat'ls £ | O & P £ | Total £ |
|---|---|---|---|---|---|---|
| Marley Ludlow Plus size 387 x 229mm | m2 | 1.04 | 14.56 | 11.76 | 3.95 | 30.27 |
| Marley Anglia Plus size 387 x 230mm | m2 | 1.04 | 14.56 | 13.97 | 4.28 | 32.81 |
| Marley Modern Plain size 420 x 330mm | m2 | 0.81 | 11.34 | 10.74 | 3.31 | 25.39 |
| Marley Modern Plain size 420 x 330mm | m2 | 0.81 | 11.34 | 10.35 | 3.25 | 24.94 |
| Marley Modern Plain size 420 x 330mm | m2 | 0.81 | 11.34 | 10.37 | 3.26 | 24.97 |
| Marley Modern Plain size 420 x 330mm | m2 | 0.81 | 11.34 | 10.02 | 3.20 | 24.56 |
| Marley Modern Plain size 420 x 330mm | m2 | 0.81 | 11.34 | 14.41 | 3.86 | 29.61 |
| Lafarge Norfolk size 381 x 227mm | m2 | 1.04 | 14.56 | 12.58 | 4.07 | 31.21 |
| Lafarge Renown size 418 x 330mm | m2 | 0.81 | 11.34 | 10.46 | 3.27 | 25.07 |
| Lafarge Double Roman size 418 x 330mm | m2 | 0.81 | 11.34 | 10.46 | 3.27 | 25.07 |
| Lafarge Regent size 418 x 332mm | m2 | 0.81 | 11.34 | 10.75 | 3.31 | 25.40 |
| Lafarge Grovebury size 418 x 332mm | m2 | 0.81 | 11.34 | 10.75 | 3.31 | 25.40 |
| Lafarge Stonewold size 430 x 380mm | m2 | 0.81 | 11.34 | 15.88 | 4.08 | 31.30 |
| fibre-cement slate size 400 x 200mm | m2 | 1.72 | 24.08 | 40.34 | 9.66 | 74.08 |
| fibre-cement slate size 500 x 250mm | m2 | 1.50 | 21.00 | 47.34 | 10.25 | 78.59 |
| fibre-cement slate size 600 x 300mm | m2 | 1.35 | 18.90 | 54.34 | 10.99 | 84.23 |
| Welsh Blue slate size 405 x 205mm | m2 | 1.80 | 25.20 | 52.85 | 11.71 | 89.76 |
| Welsh Blue slate size 510 x 255mm | m2 | 1.50 | 21.00 | 55.11 | 11.42 | 87.53 |

| | Unit | Labour Hours | £ | Mat'ls £ | O & P £ | Total £ |
|---|---|---|---|---|---|---|
| **Remove existing tiles or slates and replace with new in areas less than 1m2 (cont'd)** | | | | | | |
| Welsh Blue slate size 610 x 305mm | m2 | 1.20 | 16.80 | 58.67 | 11.32 | 86.79 |
| Westmorland slate size 450 x 230mm | m2 | 2.03 | 28.42 | 118.21 | 21.99 | 168.62 |
| Remove existing tiles or slates and replace with new in areas less than 5m2 | | | | | | |
| Marley Marlden Plain size 267 x 168mm | m2 | 1.73 | 24.22 | 33.76 | 8.70 | 66.68 |
| Marley Thaxton Plain size 270 x 168mm | m2 | 1.73 | 24.22 | 34.24 | 8.77 | 67.23 |
| Marley Heritage Plain size x 168mm | m2 | 1.73 | 24.22 | 34.50 | 8.81 | 67.53 |
| Marley Plain size 267 x 168mm | m2 | 1.73 | 24.22 | 33.40 | 8.64 | 66.26 |
| Marley Modern Plain size 420 x 330mm | m2 | 0.71 | 9.94 | 10.47 | 3.06 | 23.47 |
| Marley Ludlow Major size 267 x 168mm | m2 | 0.71 | 9.94 | 10.56 | 3.08 | 23.58 |
| Marley Ludlow Plus size 387 x 229mm | m2 | 0.83 | 11.62 | 11.76 | 3.51 | 26.89 |
| Marley Anglia Plus size 387 x 230mm | m2 | 0.83 | 11.62 | 13.97 | 3.84 | 29.43 |
| Marley Modern Plain size 420 x 330mm | m2 | 0.71 | 9.94 | 10.74 | 3.10 | 23.78 |
| Marley Modern Plain size 420 x 330mm | m2 | 0.71 | 9.94 | 10.35 | 3.04 | 23.33 |
| Marley Modern Plain size 420 x 330mm | m2 | 0.71 | 9.94 | 10.37 | 3.05 | 23.36 |

| | Unit | Labour Hours | £ | Mat'ls £ | O & P £ | Total £ |
|---|---|---|---|---|---|---|
| Marley Modern Plain size 420 x 330mm | m2 | 0.71 | 9.94 | 10.02 | 2.99 | 22.95 |
| Marley Modern Plain size 420 x 330mm | m2 | 0.71 | 9.94 | 14.41 | 3.65 | 28.00 |
| Lafarge Norfolk size 381 x 227mm | m2 | 0.83 | 11.62 | 12.58 | 3.63 | 27.83 |
| Lafarge Renown size 418 x 330mm | m2 | 0.71 | 9.94 | 10.46 | 3.06 | 23.46 |
| Lafarge Double Roman size 418 x 330mm | m2 | 0.71 | 9.94 | 10.46 | 3.06 | 23.46 |
| Lafarge Regent size 418 x 332mm | m2 | 0.71 | 9.94 | 10.75 | 3.10 | 23.79 |
| Lafarge Grovebury size 418 x 332mm | m2 | 0.71 | 9.94 | 10.75 | 3.10 | 23.79 |
| Lafarge Stonewold size 430 x 380mm | m2 | 0.71 | 9.94 | 15.88 | 3.87 | 29.69 |
| fibre-cement slate size 400 x 200mm | m2 | 1.38 | 19.32 | 40.34 | 8.95 | 68.61 |
| fibre-cement slate size 500 x 250mm | m2 | 1.20 | 16.80 | 47.34 | 9.62 | 73.76 |
| fibre-cement slate size 600 x 300mm | m2 | 1.08 | 15.12 | 54.34 | 10.42 | 79.88 |
| Welsh Blue slate size 405 x 205mm | m2 | 1.46 | 20.44 | 52.85 | 10.99 | 84.28 |
| Welsh Blue slate size 510 x 255mm | m2 | 1.20 | 16.80 | 55.11 | 10.79 | 82.70 |
| Welsh Blue slate size 610 x 305mm | m2 | 0.96 | 13.44 | 58.67 | 10.82 | 82.93 |
| Westmorland slate size 450 x 230mm | m2 | 1.62 | 22.68 | 118.21 | 21.13 | 162.02 |

| | Unit | Labour Hours | £ | Mat'ls £ | O & P £ | Total £ |
|---|---|---|---|---|---|---|
| **TIMBER SHINGLING** | | | | | | |
| Remove damaged shingles size 25 x 38 x 400mm in areas less than 1m2 and replace with new | m2 | 1.85 | 25.90 | 22.36 | 7.24 | 55.50 |
| **LEAD SHEET COVERINGS** | | | | | | |
| Clean out crack in lead roof and fill with solder | m | 1.00 | 14.00 | 4.86 | 2.83 | 21.69 |
| Take up defective lead sheeting in areas less than 1m2 and replace with new | | | | | | |
| code 5 | m2 | 3.80 | 53.20 | 24.43 | 11.64 | 89.27 |
| code 6 | m2 | 3.90 | 54.60 | 26.98 | 12.24 | 93.82 |
| Pull back existing lead flashing, dress into prepared groove and point up | m | 0.60 | 8.40 | 0.00 | 1.26 | 9.66 |
| Take off defective lead flashing and replace with new | | | | | | |
| code 5, horizontal | | | | | | |
| 150mm girth | m | 1.10 | 15.40 | 3.68 | 2.86 | 21.94 |
| 200mm girth | m | 1.20 | 16.80 | 5.86 | 3.40 | 26.06 |
| 300mm girth | m | 1.35 | 18.90 | 7.35 | 3.94 | 30.19 |
| code 5, sloping | | | | | | |
| 150mm girth | m | 1.30 | 18.20 | 3.68 | 3.28 | 25.16 |
| 200mm girth | m | 1.35 | 18.90 | 5.86 | 3.71 | 28.47 |
| 300mm girth | m | 1.40 | 19.60 | 7.35 | 4.04 | 30.99 |

| | Unit | Labour Hours | £ | Mat'ls £ | O & P £ | Total £ |
|---|---|---|---|---|---|---|
| Take off defective lead apron and replace with new | | | | | | |
| code 5, horizontal | | | | | | |
| 200mm girth | m | 1.40 | 19.60 | 5.86 | 3.82 | 29.28 |
| 300mm girth | m | 1.45 | 20.30 | 7.35 | 4.15 | 31.80 |
| 400mm girth | m | 1.50 | 21.00 | 11.00 | 4.80 | 36.80 |
| code 5, sloping | | | | | | |
| 200mm girth | m | 1.45 | 20.30 | 5.86 | 3.92 | 30.08 |
| 300mm girth | m | 1.56 | 21.84 | 7.35 | 4.38 | 33.57 |
| 400mm girth | m | 1.55 | 21.70 | 11.00 | 4.91 | 37.61 |
| Take off defective lead sill and replace with new | | | | | | |
| code 5, horizontal | | | | | | |
| 200mm girth | m | 1.30 | 18.20 | 5.86 | 3.61 | 27.67 |
| 300mm girth | m | 1.40 | 19.60 | 7.35 | 4.04 | 30.99 |
| 400mm girth | m | 1.50 | 21.00 | 11.00 | 4.80 | 36.80 |
| code 5, sloping | | | | | | |
| 200mm girth | m | 1.40 | 19.60 | 5.86 | 3.82 | 29.28 |
| 300mm girth | m | 1.50 | 21.00 | 7.35 | 4.25 | 32.60 |
| 400mm girth | m | 1.60 | 22.40 | 11.00 | 5.01 | 38.41 |
| Take off defective lead hip and replace with new | | | | | | |
| code 5, sloping | | | | | | |
| 150mm girth | m | 1.45 | 20.30 | 5.86 | 3.92 | 30.08 |
| 200mm girth | m | 1.65 | 23.10 | 7.35 | 4.57 | 35.02 |
| 300mm girth | m | 1.85 | 25.90 | 11.00 | 5.54 | 42.44 |

| | Unit | Labour Hours | £ | Mat'ls £ | O & P £ | Total £ |
|---|---|---|---|---|---|---|
| Take off defective lead kerb and replace with new | | | | | | |
| code 5, sloping | | | | | | |
| 300mm girth | m | 1.50 | 21.00 | 7.35 | 4.25 | 32.60 |
| 400mm girth | m | 1.60 | 22.40 | 11.00 | 5.01 | 38.41 |
| Take off defective lead valley and replace with new | | | | | | |
| code 5, sloping | | | | | | |
| 400mm girth | m | 1.60 | 22.40 | 11.00 | 5.01 | 38.41 |
| 600mm girth | m | 1.75 | 24.50 | 14.66 | 5.87 | 45.03 |
| Take off defective lead gutter and replace with new | | | | | | |
| code 5, sloping | | | | | | |
| 600mm girth | m | 1.75 | 24.50 | 14.66 | 5.87 | 45.03 |
| 800mm girth | m | 2.05 | 28.70 | 19.05 | 7.16 | 54.91 |
| Take off defective lead slate and replace with new | | | | | | |
| code 5 | | | | | | |
| size 400 × 400mm with 200mm high collar 100mm diameter | nr | 2.00 | 28.00 | 10.05 | 5.71 | 43.76 |
| size 400 × 400mm with 200mm high collar 150mm diameter | nr | 2.20 | 30.80 | 10.12 | 6.14 | 47.06 |

| | Unit | Labour Hours | £ | Mat'ls £ | O & P £ | Total £ |
|---|---|---|---|---|---|---|
| **COPPER SHEET COVERINGS** | | | | | | |
| Take up defective copper sheeting in areas less than 1m2 and replace with new | | | | | | |
| 0.55mm thick | m2 | 3.70 | 51.80 | 26.24 | 11.71 | 89.75 |
| 0.61mm thick | m2 | 3.70 | 51.80 | 27.64 | 11.92 | 91.36 |
| Take off defective copper flashing and replace with new | | | | | | |
| 0.55mm thick, horizontal | | | | | | |
| 150mm girth | m | 1.05 | 14.70 | 3.93 | 2.79 | 21.42 |
| 200mm girth | m | 1.15 | 16.10 | 5.25 | 3.20 | 24.55 |
| 300mm girth | m | 1.25 | 17.50 | 7.87 | 3.81 | 29.18 |
| 0.55mm thick, sloping | | | | | | |
| 150mm girth | m | 1.10 | 15.40 | 3.93 | 2.90 | 22.23 |
| 200mm girth | m | 1.20 | 16.80 | 5.25 | 3.31 | 25.36 |
| 300mm girth | m | 1.30 | 18.20 | 7.87 | 3.91 | 29.98 |
| 0.61mm thick, horizontal | | | | | | |
| 150mm girth | m | 1.05 | 14.70 | 4.15 | 2.83 | 21.68 |
| 200mm girth | m | 1.15 | 16.10 | 5.23 | 3.20 | 24.53 |
| 300mm girth | m | 1.25 | 17.50 | 8.29 | 3.87 | 29.66 |
| 0.61mm thick, sloping | | | | | | |
| 150mm girth | m | 1.10 | 15.40 | 4.15 | 2.93 | 22.48 |
| 200mm girth | m | 1.20 | 16.80 | 5.23 | 3.30 | 25.33 |
| 300mm girth | m | 1.30 | 18.20 | 8.29 | 3.97 | 30.46 |

| | Unit | Labour Hours | £ | Mat'ls £ | O & P £ | Total £ |
|---|---|---|---|---|---|---|
| Take off defective copper stepped flashing and replace with new | | | | | | |
| 0.55mm thick, sloping | | | | | | |
| 150mm girth | m | 1.20 | 16.80 | 4.43 | 3.18 | 24.41 |
| 200mm girth | m | 1.30 | 18.20 | 5.75 | 3.59 | 27.54 |
| 300mm girth | m | 1.40 | 19.60 | 8.37 | 4.20 | 32.17 |
| 0.61mm thick, sloping | | | | | | |
| 150mm girth | m | 1.20 | 16.80 | 4.65 | 3.22 | 24.67 |
| 200mm girth | m | 1.30 | 18.20 | 7.73 | 3.89 | 29.82 |
| 300mm girth | m | 1.40 | 19.60 | 8.79 | 4.26 | 32.65 |
| Take off defective copper apron and replace with new | | | | | | |
| 0.55mm thick | | | | | | |
| 200mm girth | m | 1.25 | 17.50 | 5.35 | 3.43 | 26.28 |
| 300mm girth | m | 1.30 | 18.20 | 8.02 | 3.93 | 30.15 |
| 400mm girth | m | 1.35 | 18.90 | 10.70 | 4.44 | 34.04 |
| 0.61mm thick | | | | | | |
| 200mm girth | m | 1.50 | 21.00 | 5.53 | 3.98 | 30.51 |
| 300mm girth | m | 1.55 | 21.70 | 8.29 | 4.50 | 34.49 |
| 400mm girth | m | 1.60 | 22.40 | 11.06 | 5.02 | 38.48 |
| Take off defective copper ridge capping and replace with new | | | | | | |
| 0.55mm thick | | | | | | |
| 200mm girth | m | 1.35 | 18.90 | 5.35 | 3.64 | 27.89 |
| 300mm girth | m | 1.40 | 19.60 | 8.02 | 4.14 | 31.76 |
| 400mm girth | m | 1.45 | 20.30 | 10.70 | 4.65 | 35.65 |

| | Unit | Labour Hours | £ | Mat'ls £ | O & P £ | Total £ |
|---|---|---|---|---|---|---|
| 0.61mm thick | | | | | | |
| 200mm girth | m | 1.35 | 18.90 | 5.53 | 3.66 | 28.09 |
| 300mm girth | m | 1.40 | 19.60 | 8.29 | 4.18 | 32.07 |
| 400mm girth | m | 1.45 | 20.30 | 11.06 | 4.70 | 36.06 |
| Take off defective copper hip capping and replace with new | | | | | | |
| 0.55mm thick | | | | | | |
| 200mm girth | m | 1.35 | 18.90 | 5.35 | 3.64 | 27.89 |
| 300mm girth | m | 1.40 | 19.60 | 8.02 | 4.14 | 31.76 |
| 400mm girth | m | 1.45 | 20.30 | 10.70 | 4.65 | 35.65 |
| 0.61mm thick | | | | | | |
| 200mm girth | m | 1.35 | 18.90 | 5.53 | 3.66 | 28.09 |
| 300mm girth | m | 1.40 | 19.60 | 8.29 | 4.18 | 32.07 |
| 400mm girth | m | 1.45 | 20.30 | 11.06 | 4.70 | 36.06 |
| Take off defective copper valley lining and replace with new | | | | | | |
| 0.55mm thick | | | | | | |
| 200mm girth | m | 1.35 | 18.90 | 5.35 | 3.64 | 27.89 |
| 300mm girth | m | 1.40 | 19.60 | 8.02 | 4.14 | 31.76 |
| 400mm girth | m | 1.45 | 20.30 | 10.70 | 4.65 | 35.65 |
| 0.61mm thick | | | | | | |
| 200mm girth | m | 1.35 | 18.90 | 5.53 | 3.66 | 28.09 |
| 300mm girth | m | 1.40 | 19.60 | 8.29 | 4.18 | 32.07 |
| 400mm girth | m | 1.45 | 20.30 | 11.06 | 4.70 | 36.06 |

| | Unit | Labour Hours | £ | Mat'ls £ | O & P £ | Total £ |
|---|---|---|---|---|---|---|
| **ALUMINIUM SHEET COVERINGS** | | | | | | |
| Take up defective aluminium coverings in areas less than 1m2 and replace with new | m2 | 3.50 | 49.00 | 7.48 | 8.47 | 64.95 |
| Take off defective aluminium flashing and replace with new | | | | | | |
| 0.6mm thick, horizontal | | | | | | |
| 150mm girth | m | 1.06 | 14.77 | 1.12 | 2.38 | 18.27 |
| 200mm girth | m | 1.15 | 16.10 | 1.50 | 2.64 | 20.24 |
| 300mm girth | m | 1.25 | 17.50 | 2.24 | 2.96 | 22.70 |
| 0.6mm thick, sloping | | | | | | |
| 150mm girth | m | 1.10 | 15.40 | 1.12 | 2.48 | 19.00 |
| 200mm girth | m | 1.20 | 16.80 | 1.50 | 2.75 | 21.05 |
| 300mm girth | m | 1.30 | 18.20 | 2.24 | 3.07 | 23.51 |
| 0.9mm thick, horizontal | | | | | | |
| 150mm girth | m | 1.05 | 14.70 | 1.29 | 2.40 | 18.39 |
| 200mm girth | m | 1.15 | 16.10 | 1.72 | 2.67 | 20.49 |
| 300mm girth | m | 1.25 | 17.50 | 2.58 | 3.01 | 23.09 |
| 0.9mm thick, sloping | | | | | | |
| 150mm girth | m | 1.10 | 15.40 | 1.29 | 2.50 | 19.19 |
| 200mm girth | m | 1.20 | 16.80 | 1.72 | 2.78 | 21.30 |
| 300mm girth | m | 1.30 | 18.20 | 2.58 | 3.12 | 23.90 |

| | Unit | Labour Hours | £ | Mat'ls £ | O & P £ | Total £ |
|---|---|---|---|---|---|---|
| Take off defective aluminium stepped flashing and replace with new | | | | | | |
| 0.6mm thick, sloping | | | | | | |
| 150mm girth | m | 1.20 | 16.80 | 1.12 | 2.69 | 20.61 |
| 200mm girth | m | 1.30 | 18.20 | 1.50 | 2.96 | 22.66 |
| 300mm girth | m | 1.40 | 19.60 | 2.24 | 3.28 | 25.12 |
| 0.9mm thick, sloping | | | | | | |
| 150mm girth | m | 1.20 | 16.80 | 1.29 | 2.71 | 20.80 |
| 200mm girth | m | 1.30 | 18.20 | 1.72 | 2.99 | 22.91 |
| 300mm girth | m | 1.40 | 19.60 | 2.58 | 3.33 | 25.51 |
| Take off defective aluminium apron and replace with new | | | | | | |
| 0.6mm thick | | | | | | |
| 200mm girth | m | 1.25 | 17.50 | 1.50 | 2.85 | 21.85 |
| 300mm girth | m | 1.30 | 18.20 | 2.24 | 3.07 | 23.51 |
| 400mm girth | m | 1.35 | 18.90 | 2.99 | 3.28 | 25.17 |
| 0.9mm thick | | | | | | |
| 200mm girth | m | 1.50 | 21.00 | 1.72 | 3.41 | 26.13 |
| 300mm girth | m | 1.55 | 21.70 | 2.58 | 3.64 | 27.92 |
| 400mm girth | m | 1.60 | 22.40 | 3.44 | 3.88 | 29.72 |
| Take off defective aluminium capping to ridge and replace with new | | | | | | |
| 0.6mm thick | | | | | | |
| 200mm girth | m | 1.35 | 18.90 | 1.50 | 3.06 | 23.46 |
| 300mm girth | m | 1.40 | 19.60 | 2.24 | 3.28 | 25.12 |
| 400mm girth | m | 1.45 | 20.30 | 2.99 | 3.49 | 26.78 |

| | Unit | Labour Hours | £ | Mat'ls £ | O & P £ | Total £ |
|---|---|---|---|---|---|---|
| **Take off defective aluminium capping to ridge (cont'd)** | | | | | | |
| 0.9mm thick | | | | | | |
| 200mm girth | m | 1.35 | 18.90 | 1.72 | 3.09 | 23.71 |
| 300mm girth | m | 1.40 | 19.60 | 2.58 | 3.33 | 25.51 |
| 400mm girth | m | 1.45 | 20.30 | 3.44 | 3.56 | 27.30 |
| Take off defective aluminium capping to hip and replace with new | | | | | | |
| 0.6mm thick | | | | | | |
| 200mm girth | m | 1.35 | 18.90 | 1.50 | 3.06 | 23.46 |
| 300mm girth | m | 1.40 | 19.60 | 2.24 | 3.28 | 25.12 |
| 400mm girth | m | 1.45 | 20.30 | 2.99 | 3.49 | 26.78 |
| 0.9mm thick | | | | | | |
| 200mm girth | m | 1.35 | 18.90 | 1.72 | 3.09 | 23.71 |
| 300mm girth | m | 1.40 | 19.60 | 2.58 | 3.33 | 25.51 |
| 400mm girth | m | 1.45 | 20.30 | 3.44 | 3.56 | 27.30 |
| Take off defective aluminium lining to valley and replace with new | | | | | | |
| 0.6mm thick | | | | | | |
| 200mm girth | m | 1.35 | 18.90 | 1.50 | 3.06 | 23.46 |
| 300mm girth | m | 1.40 | 19.60 | 2.24 | 3.28 | 25.12 |
| 400mm girth | m | 1.45 | 20.30 | 2.99 | 3.49 | 26.78 |
| 0.9mm thick | | | | | | |
| 200mm girth | m | 1.35 | 18.90 | 1.72 | 3.09 | 23.71 |
| 300mm girth | m | 1.40 | 19.60 | 2.58 | 3.33 | 25.51 |
| 400mm girth | m | 1.45 | 20.30 | 3.44 | 3.56 | 27.30 |

| | Unit | Labour Hours | £ | Mat'ls £ | O & P £ | Total £ |
|---|---|---|---|---|---|---|
| Take off defective aluminium lining to gutter and replace with new | | | | | | |
| 0.6mm thick | | | | | | |
| 200mm girth | m | 1.35 | 18.90 | 1.50 | 3.06 | 23.46 |
| 300mm girth | m | 1.40 | 19.60 | 2.24 | 3.28 | 25.12 |
| 400mm girth | m | 1.45 | 20.30 | 2.99 | 3.49 | 26.78 |
| 0.9mm thick | | | | | | |
| 200mm girth | m | 1.35 | 18.90 | 1.72 | 3.09 | 23.71 |
| 300mm girth | m | 1.40 | 19.60 | 2.58 | 3.33 | 25.51 |
| 400mm girth | m | 1.45 | 20.30 | 3.44 | 3.56 | 27.30 |

**ZINC SHEET COVERINGS**

| | Unit | Labour Hours | £ | Mat'ls £ | O & P £ | Total £ |
|---|---|---|---|---|---|---|
| Take up defective aluminium coverings in areas less than 1m2 and replace with new | m2 | 3.70 | 51.80 | 18.45 | 10.54 | 80.79 |
| Take off defective zinc flashing and replace with new | | | | | | |
| 0.65mm thick, horizontal | | | | | | |
| 150mm girth | m | 1.05 | 14.70 | 2.77 | 2.62 | 20.09 |
| 200mm girth | m | 1.15 | 16.10 | 3.69 | 2.97 | 22.76 |
| 300mm girth | m | 1.25 | 17.50 | 5.54 | 3.46 | 26.50 |
| 0.65mm thick, sloping | | | | | | |
| 150mm girth | m | 1.10 | 15.40 | 2.77 | 2.73 | 20.90 |
| 200mm girth | m | 1.20 | 16.80 | 3.69 | 3.07 | 23.56 |
| 300mm girth | m | 1.30 | 18.20 | 5.54 | 3.56 | 27.30 |

| | Unit | Labour | Hours £ | Mat'ls £ | O & P £ | Total £ |
|---|---|---|---|---|---|---|
| **Take off defective zinc flashing (cont'd)** | | | | | | |
| 0.80mm thick, horizontal | | | | | | |
| 150mm girth | m | 1.05 | 14.70 | 3.06 | 2.66 | 20.42 |
| 200mm girth | m | 1.15 | 16.10 | 4.07 | 3.03 | 23.20 |
| 300mm girth | m | 1.25 | 17.50 | 6.11 | 3.54 | 27.15 |
| 0.80mm thick, sloping | | | | | | |
| 150mm girth | m | 1.10 | 15.40 | 3.06 | 2.77 | 21.23 |
| 200mm girth | m | 1.20 | 16.80 | 4.07 | 3.13 | 24.00 |
| 300mm girth | m | 1.30 | 18.20 | 6.11 | 3.65 | 27.96 |
| Take off defective zinc stepped flashing and replace with new | | | | | | |
| 0.65mm thick, sloping | | | | | | |
| 150mm girth | m | 1.20 | 16.80 | 2.77 | 2.94 | 22.51 |
| 200mm girth | m | 1.30 | 18.20 | 3.69 | 3.28 | 25.17 |
| 300mm girth | m | 1.40 | 19.60 | 5.54 | 3.77 | 28.91 |
| 0.80mm thick, sloping | | | | | | |
| 150mm girth | m | 1.20 | 16.80 | 3.06 | 2.98 | 22.84 |
| 200mm girth | m | 1.30 | 18.20 | 4.07 | 3.34 | 25.61 |
| 300mm girth | m | 1.40 | 19.60 | 6.11 | 3.86 | 29.57 |
| Take off defective zinc apron and replace with new | | | | | | |
| 200mm girth | m | 1.25 | 17.50 | 3.69 | 3.18 | 24.37 |
| 300mm girth | m | 1.30 | 18.20 | 5.54 | 3.56 | 27.30 |
| 400mm girth | m | 1.35 | 18.90 | 7.38 | 3.94 | 30.22 |

| | Unit | Labour | Hours £ | Mat'ls £ | O & P £ | Total £ |
|---|---|---|---|---|---|---|
| 0.80mm thick | | | | | | |
| 200mm girth | m | 1.50 | 21.00 | 3.06 | 3.61 | 27.67 |
| 300mm girth | m | 1.55 | 21.70 | 4.07 | 3.87 | 29.64 |
| 400mm girth | m | 1.60 | 22.40 | 6.11 | 4.28 | 32.79 |
| Take off defective zinc capping to ridge and replace with new | | | | | | |
| 0.65mm thick | | | | | | |
| 200mm girth | m | 1.35 | 18.90 | 3.69 | 3.39 | 25.98 |
| 300mm girth | m | 1.40 | 19.60 | 5.54 | 3.77 | 28.91 |
| 400mm girth | m | 1.45 | 20.30 | 7.38 | 4.15 | 31.83 |
| 0.80mm thick | | | | | | |
| 200mm girth | m | 1.35 | 18.90 | 3.06 | 3.29 | 25.25 |
| 300mm girth | m | 1.40 | 19.60 | 4.07 | 3.55 | 27.22 |
| 400mm girth | m | 1.45 | 20.30 | 6.11 | 3.96 | 30.37 |
| Take off defective zinc capping to hip and replace with new | | | | | | |
| 0.65mm thick | | | | | | |
| 200mm girth | m | 1.35 | 18.90 | 3.69 | 3.39 | 25.98 |
| 300mm girth | m | 1.40 | 19.60 | 5.54 | 3.77 | 28.91 |
| 400mm girth | m | 1.45 | 20.30 | 7.38 | 4.15 | 31.83 |
| 0.80mm thick | | | | | | |
| 200mm girth | m | 1.35 | 18.90 | 3.06 | 3.29 | 25.25 |
| 300mm girth | m | 1.40 | 19.60 | 4.07 | 3.55 | 27.22 |
| 400mm girth | m | 1.45 | 20.30 | 6.11 | 3.96 | 30.37 |

| | Unit | Labour | Hours £ | Mat'ls £ | O & P £ | Total £ |
|---|---|---|---|---|---|---|
| **Take off defective zinc capping (cont'd)** | | | | | | |
| 0.65mm thick | | | | | | |
| 200mm girth | m | 1.35 | 18.90 | 3.69 | 3.39 | 25.98 |
| 300mm girth | m | 1.40 | 19.60 | 5.54 | 3.77 | 28.91 |
| 400mm girth | m | 1.45 | 20.30 | 7.38 | 4.15 | 31.83 |
| 0.80mm thick | | | | | | |
| 200mm girth | m | 1.35 | 18.90 | 3.06 | 3.29 | 25.25 |
| 300mm girth | m | 1.40 | 19.60 | 4.07 | 3.55 | 27.22 |
| 400mm girth | m | 1.45 | 20.30 | 6.11 | 3.96 | 30.37 |
| Take off defective zinc lining to valley and replace with new | | | | | | |
| 0.65mm thick | | | | | | |
| 200mm girth | m | 0.85 | 11.90 | 3.69 | 2.34 | 17.93 |
| 300mm girth | m | 0.90 | 12.60 | 5.54 | 2.72 | 20.86 |
| 400mm girth | m | 0.95 | 13.30 | 7.38 | 3.10 | 23.78 |
| 0.80mm thick | | | | | | |
| 200mm girth | m | 0.85 | 11.90 | 3.06 | 2.24 | 17.20 |
| 300mm girth | m | 0.90 | 12.60 | 4.07 | 2.50 | 19.17 |
| 400mm girth | m | 0.95 | 13.30 | 6.11 | 2.91 | 22.32 |

**BUILT-UP ROOFING**

| | Unit | Labour | Hours £ | Mat'ls £ | O & P £ | Total £ |
|---|---|---|---|---|---|---|
| Take up defective two layer built-up bituminous felt in areas less than 1m2 and and replace with new | m2 | 0.85 | 11.90 | 5.46 | 2.60 | 19.96 |

| | Unit | Labour | Hours £ | Mat'ls £ | O & P £ | Total £ |
|---|---|---|---|---|---|---|
| Take off defective skirting and replace with new | | | | | | |
| 100mm girth | m | 0.75 | 10.50 | 1.28 | 1.77 | 13.55 |
| 150mm girth | m | 0.80 | 11.20 | 1.84 | 1.96 | 15.00 |
| Take off defective flashing and replace with new | | | | | | |
| 100mm girth | m | 0.75 | 10.50 | 1.28 | 1.77 | 13.55 |
| 150mm girth | m | 0.80 | 11.20 | 1.84 | 1.96 | 15.00 |
| Take up defective three layer built-up bituminous felt in areas less than 1m2 and and replace with new | m2 | 1.00 | 14.00 | 5.46 | 2.92 | 22.38 |
| Take off defective skirting and replace with new | | | | | | |
| 100mm girth | m | 0.80 | 11.20 | 1.51 | 1.91 | 14.62 |
| 150mm girth | m | 0.85 | 11.90 | 2.01 | 2.09 | 16.00 |
| Take off defective flashing and replace with new | | | | | | |
| 100mm girth | m | 0.30 | 4.20 | 1.51 | 0.86 | 6.57 |
| 150mm girth | m | 0.35 | 4.90 | 2.01 | 1.04 | 7.95 |

# Part Four

## TOOL AND EQUIPMENT HIRE

| | 24 hours | Extra 24 hours | Week |
|---|---|---|---|
| | £ | £ | £ |

## TOOL AND EQUIPMENT HIRE

The following rates are based on average hire charges made by hire firms in the UK. Check your local dealer for more information. The rates exclude VAT

## ACCESS

### Alloy towers

| | 24 hours £ | Extra 24 hours £ | Week £ |
|---|---|---|---|
| Single width, height | | | |
| 2.2m | 41.00 | 22.00 | 82.00 |
| 3.2m | 51.00 | 26.00 | 102.00 |
| 4.2m | 60.00 | 30.00 | 120.00 |
| 5.2m | 70.00 | 35.00 | 140.00 |
| 6.2m | 78.00 | 39.00 | 156.00 |
| 7.2m | 87.00 | 43.00 | 174.00 |
| 8.2m | 96.00 | 48.00 | 192.00 |
| 9.2m | 104.00 | 52.00 | 208.00 |
| 10.2m | 112.00 | 56.00 | 224.00 |
| Full width, height | | | |
| 2.2m | 76.00 | 38.00 | 152.00 |
| 3.2m | 88.00 | 44.00 | 176.00 |
| 4.2m | 100.00 | 50.00 | 200.00 |
| 5.2m | 110.00 | 55.00 | 220.00 |
| 6.2m | 124.00 | 62.00 | 248.00 |
| 7.2m | 148.00 | 74.00 | 296.00 |
| 8.2m | 160.00 | 80.00 | 320.00 |

| | 24 hours | Extra 24 hours | Week |
|---|---|---|---|
| | £ | £ | £ |
| **Ladders** | | | |
| Roof ladders | | | |
| single, 5.9m | 24.00 | 12.00 | 48.00 |
| double, 4.6m | 22.00 | 11.00 | 44.00 |
| double, 7.6m | 26.00 | 13.00 | 46.00 |
| Push-up ladders | | | |
| double, 3.5m | 16.00 | 8.00 | 32.00 |
| double, 5.0m | 22.00 | 11.00 | 44.00 |
| treble, 2.5m | 16.00 | 8.00 | 32.00 |
| treble, 3.5m | 22.00 | 11.00 | 44.00 |
| Rope-operated ladders | | | |
| double, 6.0m | 38.00 | 19.00 | 76.00 |
| treble, 5.2m | 44.00 | 22.00 | 88.00 |
| treble, 6.0m | 50.00 | 25.00 | 100.00 |
| Stand off ladder stay | 6.00 | 3.00 | 12.00 |
| **Rubbish disposal** | | | |
| Chute, 1.2m | | | 14.00 |
| **Sundries** | | | |
| Tarpaulins | | | |
| 4 × 5m | | | 22.00 |
| 8 × 5m | | | 28.00 |
| Tar furnace | | | |
| burner | 16.00 | 8.00 | 32.00 |
| bucket | 4.00 | 2.00 | 8.00 |

| | 24 hours | Extra 24 hours | Week |
|---|---|---|---|
| | £ | £ | £ |
| Blowlamp | 10.00 | 5.00 | 20.00 |
| Flame gun | 16.00 | 8.00 | 32.00 |

# Part Five

## BUSINESS MATTERS

Starting a business

Running a business

Taxation

# Starting a business

Most small businesses come into being for one of two reasons – ambition or desperation! A person with genuine ambition for commercial success will never be completely satisfied until he has become self-employed and started his own business. But many successful businesses have been started because the proprietor was forced into this course of action because of redundancy.

Before giving up his job, the would-be businessman should consider carefully whether he has the required skills and the temperament to survive in the highly competitive self-employed market. Before commencing in business it is essential to assess the commercial viability of the intended business because it is pointless to finance a business that is not going to be commercially viable.

In the early stages it is important to make decisions such as: What exactly is the product being sold? What is the market view of that product? What steps are required before the developed product is first sold and where are those sales coming from?

As much information as possible should be obtained on how to run a business before taking the plunge. Sales targets should be set and it should be clearly established how those important first sales are obtained. Above all, do not underestimate the amount of time required to establish and finance a new business venture.

Whatever the size of the business it is important that you put in writing exactly what you are trying to do. This means preparing a business plan that will not only assist in establishing your business aims but is essential if you need to raise finance. The contents of a typical business plan are set out later. It is important to realise that you are not on your own and there are many contacts and advising agencies that can be of assistance.

### Potential customers and trade contacts

Many persons intending to start a business in the construction industry will have already had experience as employees. Use all contacts to check the market, establish the sort of work that is available and the current charge-out rates.

In the domestic market, check on the competition for prices and services provided. Study advertisements for your kind of work and try to get firm promises of work before the start-up date.

## Testing the market

Talk to as many traders as possible operating in the same field. Identify if the market is in the industrial, commercial, local government or in the domestic field. Talk to prospective customers and clients and consider how you can improve on what is being offered in terms of price, quality, speed, convenience, reliability and back-up service.

## Business links

There is no shortage of information about the many aspects of starting and running your own business. Finance, marketing, legal requirements, developing your business idea and taxation matters are all the subject of a mountain of books, pamphlets, guides and courses so it should not be necessary to pay out a lot of money for this information. Indeed, the likelihood is that the aspiring businessman will be overwhelmed with information and will need professional guidance to reduce the risk of wasting time on studying unnecessary subjects.

Business Links are now well established and provide a good place to start for both information and advice. These organisations provide a 'one-stop-shop' for advice and assistance to owner-managed businesses. They will often replace the need to contact Training and Enterprise Councils (TECs) and many of the other official organisations listed below.
Point of contact: telephone directory for address.

## Training and Enterprise Councils (TECs)

TECs are comprised of a board of directors drawn from the top men in local industry, commerce, education, trade unions etc., who, together with their staff and experienced business counsellors, assist both new and established concerns in all aspects of running a business. This takes the form of across-the-table advice and also hands-on assistance in management, marketing and finance if required. There are also training courses and seminars available in most areas together with the possibility of grants in some areas.
Point of contact: local Jobcentre or Citizens' Advice Bureau.

**Banks**

Approach banks for information about the business accounts and financial services that are available. Your local Business Link can advise on how best to find a suitable bank manager and inform you as to what the bank will require.

Shop around several banks and branches if you are not satisfied at first because managers vary widely in their views on what is a viable business proposition. Remember, most banks have useful free information packs to help business start-up.

Point of contact: local bank manager.

**HM Inspector of Taxes**

Make a preliminary visit to the local tax office enquiry counter for their publications on income tax and national insurance contributions.

| | |
|---|---|
| SA/Bk 3 | Self assessment. A guide to keeping records for the self employed |
| IR 15(CIS) | Construction Industry Tax Deduction Scheme |
| CWL | Starting your own business, |
| IR 40(CIS) | Conditions for Getting a Sub-Contractor's Tax Certificate |
| NE1 | PAYE for Employers (if you employ someone) |
| NE3 | PAYE for new and small Employers |
| IR 56/N139 | Employed or Self-Employed. A guide for tax and National Insurance |
| CA02 | National Insurance contributions for self employed people with small earnings. |

Remember, the onus is on the taxpayer, within three months, to notify the Inland Revenue that he is in business and failure to do so may result in the imposition of £100 penalty. Either send a letter or use the form provided at the back of the *'Starting your own business booklet'* to the Inland Revenue National Insurance Contributions Office and they will inform your local tax office of the change in your employment status.

Point of contact: telephone directory for address.

## Inland Revenue National Insurance Contributions Office

Self Employment Services
Customer Accounts Section
Longbenton
Newcastle NE 98 1ZZ

Telephone the Call Centre on 0845 9154655 and ask for the following publications:

| | |
|---|---|
| CWL2 | Class 2 and Class 4 Contributions for the Self Employed |
| CA02 | People with Small Earnings from Self-Employment |
| CA04 | Direct Debit - The Easy Way to Pay. Class 2 and Class 3 |
| CA07 | Unpaid and Late Paid Contributions and for Employers |
| CWG1 | Employer's Quick Guide to PAYE and NIC Contributions |
| CA30 | Employer's Manual to Statutory Sick Pay |

## VAT

The VAT office also offer a number of useful publications, including;

| | |
|---|---|
| 700 | The VAT Guide |
| 700/1 | Should I be Registered for VAT? |
| 731 | Cash Accounting |
| 732 | Annual Accounting |
| 742 | Land and Property |

Information about the Cash Accounting Scheme and the introduction of annual VAT returns are dealt with later.
Point of contact: telephone directory for address.

## Local authorities

Authorities vary in provisions made for small businesses but all have been asked to simplify and cut delays in planning applications. In Assisted Areas, rent-free periods and reductions in rates may be available on certain industrial and commercial properties. As a preliminary to either purchasing or renting business premises, the following booklets will be helpful:

*Step by Step Guide to Planning Permission for Small Businesses,* and *Business Leases and Security of Tenure*

Both are issued by the Department of Employment and are available at council offices, Citizens' Advice Bureau and TEC offices. Some authorities run training schemes in conjunction with local industry and educational establishments.

Point of contact: usually the Planning Department - ask for the Industrial Development or Economic Development Officer.

**Department of Trade and Industry**

The services formally provided by the Department are now increasingly being provided by Business Link . The Department can still, however, provide useful information on available grants for start-ups.
Point of contact: telephone 0207-215 5000 and ask for the address and telephone number of the nearest DTI office and copies of their explanatory booklets.

**Department of Transport and the Regions**

Regulations are now in force relating to all forms of waste other than normal household rubbish. Any business that produces, stores, treats, processes, transports, recycles or disposes of such waste has a 'duty of care' to ensure it is properly discarded and dealt with.

Practical guidance on how to comply with the law (it is a criminal offence punishable by a fine not to) is contained in a booklet *Waste Management: The Duty of Care: A Code of Practice,* obtainable from HMSO Publication Centre, PO Box 276, London SW8 5DT. Telephone 0207-873 9090.

**Accountant**

The services of an accountant are to be strongly recommended from the beginning because the legal and taxation requirements start immediately and must be properly complied with if trouble is to be avoided later. A qualified accountant must be used if a limited company is being formed but an accountant will give advice on a whole range of business issues including book-keeping, tax planning and compliance to finance raising and will help in preparing annual accounts.

It is worth spending some time finding an accountant who has other clients in the same line of business and is able to give sound advice particularly on taxation and business finance and is not so overworked that damaging delays in producing accounts are likely to arise. Ask other traders whether they can recommend their own accountant. Visit more than one firm of accountants, ask about the fees they charge and how much the production of annual accounts and agreement with the Inland Revenue are likely to cost. A good accountant is worth every penny of his fees and will save you both money and worry.

## Solicitor

Many businesses operate without the services of a solicitor but there are a number of occasions when legal advice should be sought. In particular, no-one should sign a lease of premises without taking legal advice because a business can encounter financial difficulty through unnoticed liabilities in its lease. Either an accountant or solicitor will help with drawing up a partnership agreement that all partnerships should have. A solicitor will also help to explain complex contractual terms and prepare draft contracts if the type of business being entered into requires them.

## Insurance broker

Policies are available to cover many aspects of business including:

- employer's liability - compulsory if the business has employees
- public liability - essential in the construction industry
- motor vehicles
- theft of stock, plant and money
- fire and storm damage
- personal accident and loss of profits
- key man cover.

Brokers are independent advisers who will obtain competitive quotations on your behalf. See more than one broker before making a decision - their advice is normally given free and without obligation.
Point of contact: telephone directory or write for a list of local members to:

The British Insurance Brokers' Association
Consumer Relations Department
BIBA House
14 Bevis Marks
London
EC3A 7NT (telephone: 0207-623 9043)

or contact
The Association of British Insurers
51 Gresham Street
London
EC2V 7HQ (telephone: 0207-600 3333)

who will supply free a package of very useful advice files specially designed for the small business.

**The Health and Safety Executive**

The Executive operates the legislation covering everyone engaged in work activities and has issued a very useful set of '*Construction Health Hazard Information Sheets*' covering such topics as handling cement, lead and solvents, safety in the use of ladders, scaffolding, hoists, cranes, flammable liquids, asbestos, roofs and compressed gases etc. A pack of these may be obtained free from your local HSE office or The Health & Safety Executive Central Office, Sheffield (telephone: (01142-892345) or HSE Publications (telephone: 01787-881165).

**Business plan**

As stated before, once the relevant information has been obtained it should be consolidated into a formal business plan. The complexity of the plan will depend in the main on the size and nature of the business concerned. Consideration should be given to the following points.

Objectives

It is important to establish what you are trying to achieve both for you and the business. A provider of finance may be particularly influenced by your ability to achieve short- and medium-term goals and may have confidence in continuing to provide finance for the business. From an individual point of view, it is important to establish goals because there is little point in having a business that only serves to achieve the expectations of others whilst not rewarding the would-be businessman.

History

If you already own an existing business then commentary on its existing background structure and history to date can be of assistance. There is no substitute for experience and any existing contacts you have in the construction industry will be of assistance to you. The following points should also be considered for inclusion:

- a brief history of the business identifying useful contacts made
- the development of the business, highlighting significant successes and their relevance to the future
- principal reasons for taking the decision to pursue this new venture
- details of present financing of the business.

Products or services

It is important to establish precisely what it is you are going to sell. Does the product or service have any unique qualities which gives it your advantages over competitors? For example, do you have an ability to react more quickly than your competitors and are you perceived to deliver a higher quality product or service? A typical business plan would include:

- description of the main products and services
- statement of disadvantages and advising how they will be overcome
- details of new products and estimated dates of introduction

- profitability of each product
- details of research and development projects
- after-sales support.

Markets and marketing strategy

This section of the business plan should show that thought has been given to the potential of the product. In this regard it can often be useful to identify major competitors and make an overall assessment of their strengths and weaknesses, including the following:

- an overall assessment of the market, setting out its size and growth potential
- a statement showing your position within the market
- an identification of main customers and how they compare
- details of typical orders and buying habits
- pricing strategy
- anticipated effect on demand of pricing
- expectation of price movement
- details of promotions and advertising campaigns.

It is important to identify your customers and why they might buy from you. Those entering the domestic side of the business will need to think about the best way to reach potential customers. Are local word-of-mouth recommendations enough to provide reasonable work continuity. If not, what is the most effective method of advertising to reach your customer base?

Remember, advertising is costly. It is a waste of funds to place an advertisement in a paper circulating in areas A, B, C & D if the business only covers area A.

Research and development

If you are developing a product or a particular service, then an assessment should be made on what stage it is at and what further finance is required to complete it. It may also be useful to make an assessment on the vulnerability of the product or service to innovations being initiated by others.

Basis of operation

Detail what facilities you will require in order to carry on your trade in the form of property, working and storage areas, office space, etc. An assessment should also he made on the assistance you will require from others. Your business plan might include:

- a layman's guide to the process or work
- details of facilities, buildings and plant
- key factors affecting production, such as yields and wastage
- raw material demand and usage.

Management

This section is one of the most important because it demonstrates the capability of the would-be businessman. The skills you need will cover production, marketing, finance and administration. In the early stages you may be able to do this yourself but as the business grows it may be required to develop a team to handle these matters. The following points should be considered for inclusion in the plan:

- set out age, experience and achievements
- state additional management requirements in the future and how they are to be met
- identify current weaknesses and how they will be overcome
- state remuneration packages and profit expectations
- give detailed CVs in appendices.

Advertising and retraining may be required in order to identify and provide suitable personnel where expertise and experience are lacking.

Financial information

It is important to detail, if any, the present financial position of your business and the budgeted profit and loss accounts, cash flows and balance sheets. These integrated forecasts should be prepared for the next twelve months at monthly intervals and annually for the following two years.

If the forecasts are to be reasonably accurate then the businessman must make some early decisions about:

- the premises where the business will be based, the initial repairs and alterations that might he required and an assessment of the total cost
- which plant, equipment and transport are needed, whether they are to be leased or purchased and what the cost will be?
- how much stock of materials, if any, should be carried? - the bare minimum only should be acquired, so reliable suppliers should be found
- what will be the weekly bills for overheads, wages and the proprietor's living costs?
- what type of work is going to be undertaken, and how much profit can realistically be obtained?
- how often are invoices to be presented?

Your business plan should include the following information:

- explanation of how sales forecasts are prepared
- levels of production
- details of major variable overheads and estimates
- assumptions in cash flow forecasting, inflation and taxation.

Finance required and its application

The financial details given above should produce an accurate assessment of the funds required to finance the business. It is important to distinguish between those items that require permanent finance and those that will eventually be converted to cash because it is not usually advisable to finance long-term assets with personal equity.

Working capital such as stock and debtors can usually be obtained by an overdraft arrangement but your accountant or bank will advise you on this.

Executive summary

Although it is prepared last, this summary will be the first part of your business plan. Remember that business plans are prepared for busy people and their decision on finance may be based solely on this section. It should cover two or three pages and deal with the most important aspects and opportunities in your plan. Here are some of the main headings:

- key strategies
- finance required and how it is to be used
- management experience
- anticipated returns and profits
- markets.

The appendices should include:

- CVs of key personnel
- organisation charts
- market studies
- product advertising literature
- professional references
- financial forecasts
- glossary of terms.

If you feel that any additional information should be provided in support of your proposal, then this is usually best included in the appendices.

Follow up

Please remember that once your plan is prepared, it is important to re-examine it regularly and update the forecasts and financial information. This is a working document and can be an important tool in running the business.

**Sources of finance**

Personal funds

Finance, like charity, often begins at home and a would-be businessman should make a realistic assessment of his net worth, including the value of his house after deducting the mortgage(s) outstanding on it, savings, any car or van owned and any sums which the family are prepared to contribute but deducting any private borrowings which will come due for payment. The whole of these funds may not be available (for instance, money which has been loaned to a friend or relative who is known to be unable to repay at the present time).

It may not be desirable that all capital should be put at risk on a business venture so the following should be established:

- how much cash you propose to invest in the business
- whether the family home will be made available for any business borrowing
- state total finance required
- how finance is anticipated being raised
- interest and security to be provided
- expected return on investment.

Whilst it may be wise not to pledge too much of the family assets, it has to be remembered that the bank will be looking closely at the degree to which the proprietor has committed himself to the venture and will not be impressed by an application for a loan where the applicant is prepared to risk only a small fraction of his own resources.

Having decided how much of his own funds to contribute, the businessman can now see the level of shortfall and consider how best to fill it. Consideration should be given to partners where the shortfall is large and particularly when there is a need for heavy investment in fixed assets, such as premises and capital equipment. It may be worthwhile starting a limited company with others also subscribing capital and to allow the banks to take security against the book debts.

Banks

The first outside source of money to which most businessmen turn is the bank and here are a few guidelines on approaching a bank manager:

- present your business plan to him; remember to use conservative estimates which tend to understate rather than overstate the forecast sales and profits
- know the figures in detail and do not leave it to your accountant to explain them for you. The bank manager is interested in the businessman not his advisers and will be impressed if the businessman demonstrates a grasp of the financing of his business

- understand the difference between short- and long-term borrowing
- ask about the Government Loan Guarantee Scheme if there is a shortage of security for loans. The bank may be able to assist, or depending on certain conditions being met, the Government may guarantee a certain percentage of the bank loan.

Remember the bank will want their money back, so bank borrowings are usually required to be secured by charges on business assets. In start-up situations, personal guarantees from the proprietors are normally required. Ensure that if these are given they are regularly reviewed to see if they are still required.

Enterprise Investment Scheme - business angels

If an outside investor is sought in a business he will probably wish to invest within the terms of the Enterprise Investment Scheme which enables him to gain income tax relief at 20% on the amount of his investment. Additionally, any investment can be used to defer capital gains tax. The rules are complex and professional advice should always be sought.

Hire purchase/leasing

It is not always necessary to purchase assets outright that are required for the business and leasing and hire purchase can often form an integral part of a business's medium-term finance strategy.

Venture capital

In addition, there are a number of other financial institutions in the venture capital market that can help well-established businesses, usually limited companies, who wish to expand. They may also assist well-conceived start-ups. They will provide a flexible package of equity and loan capital but only for large amounts, usually sums in excess of £150,000 and often £250,000.

Usually the deal involves the financial institution having a minority interest in the voting share capital and a seat on the board of the company. Arrangements for the eventual purchase of the shares held by the finance company by the private shareholders are also normally incorporated in the scheme.

The Royal Jubilee and Princes Trust

These trusts through the Youth Business Initiative provide bursaries of not more than £1,000 per individual to selected applicants who are unemployed and age 25 or over. Grants may be used for tools and equipment, transport, fees, insurance, instruction and training but not for working capital, rent and rates, new materials or stock. They operate through a local representative whose name and address may be ascertained by contacting the Prince's Youth Business.
Point of contact: telephone 0207-321 6500.

The Business Start-up Scheme

This is an allowance of £50 per week, in addition to any income made from your business, paid for twenty weeks. To qualify you must be at least 18 and under 65, work at least 36 hours per week in the business and have been unemployed for at least six months or fall into one of the other categories: disabled, ex-HMS or redundant.

The first step is to get the booklet on the subject from your local Jobcentre or TEC that includes details on how and where to apply. Once in receipt of the enterprise allowance, you will also have the benefit of advice and assistance from an experienced businessman from your TEC. All the initial counselling services and training courses are free.

# Running a business

Many businesses are run without adequate information being available to check trend in their vital areas, e.g. marketing, money and managerial efficiency. It is essential to look critically at all aspects of the business in order to maximise profits and reduce inefficiency. Regular meaningful information is required on which management can concentrate. This will vary according to the proprietor's business but will often concentrate on debtors, creditors, cash, sales and orders.

Proprietors often have the feeling that the business should be 'doing better' but are unable to identify what is going wrong. Sometimes there is the worrying phenomenon of a steadily increasing work programme coupled with a persistently reducing bank balance or rising overdraft. Some useful ways of checking the position and of identifying problem areas are given below.

## Marketing

Throughout his business life the entrepreneur should continuously study the methods and approach of his competitors. A shortcoming frequently found in ailing concerns is that the proprietor thinks he knows what his customers want better than they do.

The term 'market research' sounds both difficult and expensive but a very simple form of it can be done quite effectively by the businessman and his sales staff. Existing and prospective customers should be approached and asked what they want in terms of price, quality, design, payment terms, follow-up service, guarantees and services.

The initial approach might be by a leaflet or letter followed by a personal call. As an on-going part of management, all staff with customer contact should be encouraged to enquire about and record customer preferences, complaints, etc. and feed it back to management.

Other sources of information can be trade and business journals, trade exhibitions, suppliers and representatives from which information about trends, new techniques and products can be obtained and studied. Valuable information can also be gained from studying competitors and the following questions should be asked:

- what do they sell and at what prices?
- what inducements do they offer to their customers, e.g. credit facilities, guarantees, free offers and discounts?
- how do they reach their customers - local/national advertising, mail shots, salesmen, local radio and TV?
- what are the strongest aspects of their appeal to customers and have they any weaknesses?

The businessman should apply all the information gathered from customers and competitors to his own services with a view to making sure he is offering the right product at the right price in the most attractive way and in the most receptive market.

In a small business where the proprietor is also his own salesman he must give careful thought on how he can best present his product and himself. For instance, if he is working solely within the construction industry his main problems are likely to centre on getting a C1S6 Certificate and using trade contacts to get sub-contract work.

However, for those who serve the general public, presentation can be a vital element in getting work. The customer is looking for efficiency, reliability and honesty in a trader and quality, price and style in the product. To bring out these facets in discussion with a potential customer is a skilled task. A short course on marketing techniques could pay handsome dividends. The Business Link will give the names and addresses of such courses locally.

**Financial control**

Unfortunately, some unsuccessful firms do not seek financial advice until too late when the downward trend cannot be halted. Earlier attention to the problems may have saved some of them so it is important to recognise the tell-tale signs. There are some tests and checks that can be done quite easily.

**Cash flow**

Cash flow is the lifeblood of the business and more businesses fail through lack of cash than for any other reason. Cash is generated through the conversion of work into debtors and then into payment and also through the deferral of the payment of supplies for as long a period that can be negotiated. The objective must be to keep stock, work in progress, debts to a minimum and creditors to a maximum.

### Debtor days

This is calculated by dividing your trade debtors by annual sales and multiplying by 365. This shows the number of days' credit being afforded to your customers and should be compared both with your normal trade terms and the previous month's figures. Normal procedures should involve the preparation of a monthly-aged list of debtors showing the name of the customer, the value and to which month it relates.

The oldest and largest debtors can be seen at a glance for immediate consideration of what further recovery action is needed. The list may also show over-reliance on one or two large customers or the need to stop supplying a particularly bad payer until his arrears have been reduced to an acceptable level. Consideration should be given to making up bills to a date before the end of the month and making sure the accounts are sent out immediately, followed by a statement four weeks later.

Consider giving discounts for prompt payment. If all else fails, and legal action for recovery is being contemplated, call at the County Court and ask for their leaflets.

### Stock turn

The level of stock should be kept to a minimum and the number of days' stock can be calculated by dividing the stock by the annual purchases and multiplying by 365. A worsening trend on a month-by-month basis shows the need for action. It is important to regularly make a full inventory of all stock and dispose of old or surplus items for cash. A stock control procedure to avoid stock losses and to keep stock to a minimum should be implemented.

### Profitability

Whilst cash is vital in the short-term, profitability is vital in the medium-term. The two key percentage figures are the gross profit percentage and the net profit percentage. Gross profit is calculated by deducting the cost of materials and direct labour from the sales figures whilst net profit is arrived at after deducting all overheads. Possible reasons for changes in the gross profit percentage are:

- not taking full account of increases in materials and wages in the pricing of jobs
- too generous discount terms being offered
- poor management, over-manning, waste and pilferage of materials
- too much down-time on equipment which is in need of replacement.

If net profit is deteriorating after the deduction of an appropriate reward for your own efforts, including an amount for your own personal tax liability, you should review each item of overhead expenditure in detail asking the following questions:

- can savings be made in non-productive staff?
- is sub-contracting possible and would it be cheaper?
- have all possible energy-saving methods been fully explored?
- do the company's vehicles spend too much time in the yard and can they be shared or their number reduced?
- is the expenditure on advertising producing sales - review in association with 'marketing' above?

## Over-trading

Many inexperienced businessmen imagine that profitability equals money in the bank and in some cases, particularly where the receipts are wholly in cash, this may be the case. But often, increased business means higher stock inventories, extra wages and overheads, increased capital expenditure on premises and plant, all of which require short-term finance.

Additionally, if the debtors show a marked increase as the turnover rises, the proprietor may find to his surprise that each expansion of trade reduces rather than increases his cash resources and he is continually having to rely on extensions to his existing credit.

The business, which had enough funds for start-up, finds it does not have sufficient cash to run at the higher level of operation and the bank manager may he getting anxious about the increasing overdraft. It is essential for those who run a business that operates on credit terms to be aware that profitability does not necessarily mean increased cash availability. Regular monthly management information on marketing and finance as described in this chapter will enable over-trading to be recognised and remedial action to be taken early.

If the situation is appreciated only when the bank and other creditors are pressing for money, radical solutions may be necessary, such as bringing in new finance, sale and leaseback of premises, a fundamental change in the terms of trade or even selling out to a buyer with more resources. Help from the firm's accountant will be needed in these circumstances.

## Break-even point

The costs of a business may be divided into two types - variable and fixed. *Variable costs* are those which increase or decrease as the volume of work goes up or down and include such items as materials used, direct labour and power machine tools. *Fixed costs* are not related to turnover and are sometimes called fixed overheads. They include rent, rates, insurance, heat and light, office salaries and plant depreciation. These costs are still incurred even though few or no sales are being made.

Many small businessmen run their enterprises from home using family labour as back-up; they mainly sell their own labour and buy materials and hire plant only as required. By these means they reduce their fixed costs to a minimum and start making profits almost immediately. However, larger firms that have business premises, perhaps a small workshop, an office and vehicles, need to know how much they have to sell to cover their costs and become profitable.

In the case of a new business it is necessary to estimate this figure but where annual accounts are available a break-even chart based on them can be readily prepared. Suppose the real or estimated figures (expressed in £000s) are:

| | % | £ |
|---|---|---|
| Sales | 100 | 400 |
| Variable costs | 66 | 265 |
| Gross profit | 34 | 135 |
| Fixed costs | 13 | 50 |
| Net profit | 21 | 85 |

$$\text{Break-even point} = \frac{\text{50 divided by (1 less variable costs \%)}}{\text{sales}}$$

= 50 divided by (1 less 0.6625)
= 50 divided by 0.3375
= £148 (thousand)

In practice, things are never quite as clear cut as the figures show, but nevertheless this is a very useful tool for assessing not only the break-even point but also the approximate amount of loss or profit arising at differing levels of turnover and also for considering pricing policy.

## Taxation

The first decision usually required to be made from a taxation point of view is which trading entity to adopt. The options available are set out below.

### Sole trader

A sole trader is a person who is in business on his own account. There is no statutory requirement to produce accounts nor is there a necessity to have them audited. A sole trader may, however, be required to register for PAYE and VAT purposes and maintain records so that Income Tax and VAT returns can be made. A sole trader is personally liable for all the liabilities of his business.

### Partnership

A partnership is a collection of individuals in business on their own account and whose constitution is generally governed by the Partnership Act 1890. It is strongly recommended that a partnership agreement is also established to determine the commercial relationship between the individuals concerned.

The requirements in relation to accounting records and returns are similar to those of a sole trader and in general a partner's liability is unlimited.

### Limited company

This is the most common business entity. Companies are incorporated under the Companies Act 1985 which requires that an annual audit is carried out for all companies with a turnover in excess of £5,000,000 or a review if the turnover is less than £5,000,000 and that accounts are filed with the Companies Registrar. Generally an individual shareholder's liability is limited to the amount of the share capital he is required to subscribe.

### Advantages

In view of the problems and costs of incorporating an existing business, it is important to try and select the correct trading medium at the commencement of operations. It is not true to say that every business should start life as a company.

Many businesses are carried on in a safe and efficient manner by sole traders or partnerships. Whilst recognising the possible commercial advantages of a limited company, taxation advantages exist for sole traderships and partnerships, such as income tax deferral and National Insurance saving. No decision should be taken without first seeking professional advice.

The benefit of limited liability should not be ignored although this can largely be negated by banks seeking personal guarantees. In addition, it may be easier for the companies to raise finance because the bank can take security on the debts of the company that could be sold in the future, particularly if third-party finance has been obtained in the form of equity.

**Self-assessment**

From the tax year 1996/97 the burden of assessing tax shifted from the Inland Revenue to the individual tax payer. The main features of this system are as follows:

- the onus is on the taxpayer to provide information and to complete returns
- tax will be payable on different dates
- the taxpayer has a choice: he can calculate his tax liability at the same time as making his return and this will need to be done by 31st January following the end of the tax year. Alternatively, he can send in his tax return before 30 September and the Inland Revenue will calculate the tax to be paid on the following 31 January
- the important aspect to the system is that if the return is late, or the tax is paid late, there will be automatic penalties and/or surcharges imposed on the taxpayer.

**Tax correspondence**

Businessmen do not like letters from the Inland Revenue but they should resist the temptation to tear them up or put them behind the clock and forget about them. All Tax Calculations and Statements of Account should be checked for accuracy immediately and any queries should be put to your accountant or sent to the Tax District that issued the document.

Keep copies of all correspondence with the Inland Revenue. Letters can be mislaid or fail to be delivered and it is essential to have both proof of what was sent as well as a permanent record of all correspondence.

## Dates tax due

*Income Tax*

Payments on account (based on one half of last year's liability) are due on 31 January and 31 July. If these are insufficient there is a balancing payment due on the following 31 January – the same day as the tax return needs to be filed. For example:

| | |
|---|---|
| for the year 2004/05 | Tax due £5,000 (2003/04 was £4,000) |
| | First payment on account of £2,000 is due on 31.01.05 |
| | Second payment on account of £2,000 is due on 31.07.05 |
| | Balancing payment of £1,000 is due on 31.01.06 |

Note that on 31.01.06, the first payment on account of £2,500 fell due for the tax year 2005/06.

## Tax in business

*Spouses in business*

If spouses work in the business, perhaps answering the phone, making appointments, writing business letters, making up bills and keeping the books, they should be properly remunerated for it. Being a payment to a family member, the Inspector of Taxes will be understandably cautious in allowing remuneration in full as a business expense. The payment should be:

- actually paid to them, preferably weekly or monthly and in addition to any housekeeping monies
- recorded in the business book

- reasonable in amount in line with their duties and the time spent on them.

If the wages paid to them exceed £91.00 per week, Class 1 employer's and employee's NIC becomes due and if they exceed £4,745 p.a. (assuming they have no other income) PAYE tax will also be payable.

It should also be noted that once small businesses are well established and the spouses' earnings are approaching the above limits, consideration may be given to bringing them in as a partner. This has a number of effects:

- there is a reduced need to relate the spouse's income (which is now a share of the profits) to the work they do
- they will pay Class 2 and Class 4 NIC instead of the more costly Class I contributions and PAYE will no longer apply to their earnings but remember that, as partners, they have unlimited liability.

**Premises**

Many small businessmen cannot afford to rent or buy commercial premises and run their enterprises from home using part of it as an office where the books and vouchers, clients' records and trade manuals are kept and where estimates and plans are drawn up. In these circumstances, a portion of the outgoings on the property may be claimed as a business expenses. An accountant's advice should be sought to ensure that the capital gains tax exemption that applies on the sale of the main residence is not lost.

**Fixed Profit Car Scheme**

It may be advantageous to calculate your car expenses using a fixed rate per business mile. A condition is that your annual turnover is below the VAT threshold (currently £58,000). Ask your accountant about this. A proper record of business mileage must be kept.

**Vehicles**

Car expenses for sole traders and partners are usually split on a fractional mileage basis between business journeys, which are allowable, and private ones, which are not, and a record of each should he kept. If the business does work only on one or two sites for only one main contractor, the inspector may argue that the true base of operations is the work site not the residence and seek to disallow the cost of travel between home and work. It is tax-wise and sound business practice to have as many customers as possible and not work for just one client.

**Business entertainment**

No tax relief is due for expenditure on business entertainment and neither is the VAT recoverable on gifts to customers, whether they are from this country or overseas. However, the cost of small trade gifts not exceeding £50 per person per annum in value is still admissible provided that the gift advertises the business and does not consist of food, drink or tobacco.

**Income tax (2004/05)**

*Personal allowances*

The current personal allowance for a single person is £4,745. The personal allowance for people aged 65 to 74 and over 75 years are £6,830 and £6,950 respectively. The married couple's allowance was withdrawn on 5 April 2000, except for those over 65 on that date.

*Taxation of husband and wife*

A married woman is treated in much the same way as a single person with her own personal allowance and basic rate band. Husband and wife each make a separate return of their own income and the Inland Revenue deals with each one in complete privacy; letters about the husband's affairs will be addressed only to him and about the wife's only to her unless the parties indicate differently.

**Rates of tax**

Tax is deducted at source from most banks and building societies accounts at the rate of 20%. The rates of tax for 2004/05 are as follows:

Lower rate: 10% on taxable income up to £2,020
Basic rate: 22% on taxable income between £2,020 and £31,400
Higher rate: 40% on taxable income over £31,400

Dividends carry a 10% non-repayable tax credit. Higher rate taxpayers pay a further tax on dividends of 22.5%.

### Mortgage interest relief

This is no longer available after 5 April 2000.

### Business losses

These are allowed only against the income of the person who incurs the loss. For example, a loss in the husband's business cannot be set against the wife's income from employment.

### Joint income

In the case of joint ownership by a husband and wife of assets that yield income, such as bank and building society accounts, shares and rented property, the Inland Revenue will treat the income as arising equally to both and each will pay tax on one half of the income. If, however, the asset is owned in unequal shares or one spouse only and the taxpayer can prove this, then the shares of income to be taxed can be adjusted accordingly if a joint declaration is made to the tax office setting out the facts.

### Capital Gains Tax

Where an asset is disposed of, the first £8,200 of the gain is exempt from tax. In the case of husbands and wives, each has a £8,200 exemption so if the ownership of the assets is divided between them, it is possible to claim exemption on gains up to £16,400 jointly in the tax year. Any remaining gain is chargeable as though it were the top slice of the individual's income; therefore according to his or her circumstances it might be charged at 10%, 22% or 40%.

**Self-employed NIC rates (from 6 April 2004)**

*Class 2 rate*
Charged at £2.00 per week. If earnings are below £4,745 per annum averaged over the year, ask the DSS about 'small income exception'. Details are in leaflet CA02.

*Class 4 rate*
Business profits up to £4,745 per annum are charged at NIL. Annual profits between £4,615 and £31,720 are charged at 8% of the profit. There is also a charge on profits over £31,720 of 1%. Class 4 contributions are collected by the Inland Revenue along with the income tax due.

*Capital allowances (depreciation) rates*

| | |
|---|---|
| Plant and machinery: | 25% (40% first-year allowance is available for certain small businesses) |
| Business motor cars - cost up to £12,000: | 25% |
| - cost over £12,000: | £3,000 (maximum) |
| Industrial build | 4% |
| Commercial and industrial buildings in Enterprise Zones: | 100% |
| Computers and software equipment | 50% |

THE CONSTRUCTION INDUSTRY TAX DEDUCTION SCHEME

**General**

The new Construction Industry Tax Deduction Scheme is known as the 'CIS' scheme and replaced the old '714' scheme. As the scheme operates whenever a contractor makes a payment to a sub-contractor, the businessman should visit his local income tax enquiry office and obtain copies of the Inland Revenue booklet IR 14/15 (CIS) and leaflet IR 40 which will explain the conditions under which the Inland Revenue will issue a registration card or (CIS6) certificate and precisely when the scheme applies.

Everyone who carries out work in the Construction Industry Scheme must hold a registration card (CIS4) or a tax certificate (CIS6). Certain larger companies use a special certificate (CIS5).

If the sub-contractor has a registration card but does not hold a valid tax certificate (CIS6) issued to him by the Inland Revenue, then the contractor *must* deduct 18% tax from the whole of any payment made to him (excluding the cost of any materials) and to account to the Inland Revenue for all amounts so withheld.

To enable the subcontractor to prove to the Inspector of Taxes that he has suffered this tax deduction, the contractor must complete the three-part tax payment voucher (CIS25) showing the amount withheld. These vouchers must be carefully filed for production to the Inspector after the end of the tax year along with the tax return. Any tax deducted in this way over and above the sub-contractor's agreed liability for the year will be repaid by the Inland Revenue. If he holds a (CIS6) certificate the payment may be made in full without deducting tax.

A small business that does work only for the general public and small commercial concerns is outside the scheme and does not need a certificate to trade. If, however, it engages other contractors to do jobs for it, the business would have to register under the scheme as a contractor and deduct tax from any payment made to a sub-contractor who did not produce a valid (CIS6) certificate. If in doubt, consult your accountant or the Inland Revenue direct.

## VAT

The general rule about liability to register for VAT is given in the VAT office notes. It is possible to give here only a brief outline of how the tax works. The rules that apply to the construction industry are extremely complex and all traders must study *The VAT Guide* and other publications.

Registration for VAT is required if, at the end of any month, the value of taxable supplies in the last 12 months exceeds the annual threshold or if there are reasonable grounds for believing the value of the taxable supplies in the next 30 days will exceed the annual threshold.

Taxable supplies include any zero-rated items. The annual threshold is £58,000. The amount of tax to be paid is the difference between the VAT charged out to customers *(output tax)* and that suffered on payments made to suppliers for goods and services *(input tax)* incurred in making taxable supplies. Unlike income tax there is no distinction in VAT for capital items so that the tax charged on the purchase of, for example, machinery, trucks and office furniture, will normally be reclaimable as *input tax*.

VAT is payable in respect of three monthly periods known as 'tax periods'. You can apply to have the group of tax periods that fits in best with your financial year. The tax must be paid within one month of the end of each tax period. Traders who receive regular repayments of VAT can apply to have them monthly rather than quarterly. Not all types of goods and services are taxed at 17.5% (i.e. the standard rate). Some are exempt and others are zero-rated.

**Zero-rated**

This means that no VAT is chargeable on the goods or services, but a registered trader can reclaim any *input* tax suffered on his purchases. For instance, a builder pays VAT on the materials he buys to provide supplies of constructing but if he is constructing a new dwelling house, this is zero rated. The builder may reclaim this VAT or set it off against any VAT due on standard rated work.

**Exempt**

Supplies that are exempt are less favourably treated than those that are zero rated. Again no VAT is chargeable on the goods or services but the trader cannot reclaim any *input* tax suffered on his purchases.

**Standard-rated**

All work which is not specifically stated to be zero rated or exempt is standard-rated, i.e. VAT is chargeable at the current rate of 17.5% and the trader may deduct any *input* tax suffered when he is making his return to the Customs and Excise. If for any reason a trader makes a supply and fails to charge VAT when he should have done so (e.g. mistakenly assuming the supply to be zero rated), he will have to account for the VAT himself out of the proceeds. If there is any doubt about the VAT position, it is safer to assume the supply is standard rated, charge the appropriate amount of VAT on the invoice and argue about it later.

## Time of supply

The *time* at which a supply of goods or services is treated as taking place is important and is called the 'tax point'. VAT must be accounted for to the Customs and Excise at the end of the accounting period in which this 'tax point' occurs. For the supply of goods which are 'built on site', the 'basic tax point' is the date the goods are made available for the customer's use, whilst for *services* it is normally the date when all work except invoicing is completed.

However, if you issue a tax invoice or receive a payment before this 'basic tax point' then that date becomes a tax point. In the case of contracts providing for stage and retention payments, the tax point is either the date the tax invoice is issued or when payment is received, whichever is the earlier.

All the requirements apply to sub-contractors and main contractors and it should be noted that, when a contractor deducts income tax from a payment to a sub-contractor (because he has no valid CIS6) VAT is payable on the full gross amount *before* taking off the income tax.

## Annual accounting

It is possible to account for VAT other than on a specified three month period. Annual accounting provides for nine equal installments to be paid by direct debit with annual return provided with the tenth payment. £300,000.

## Cash accounting

If turnover is below a specified limit, currently £660,000, a taxpayer may account for VAT on the basis of cash paid and received. The main advantages are automatic bad debt relief and a deferral of VAT payment where extended credit is given.

## Bad debts

Relief is available for debts over 6 months.

# Part Six

## GENERAL CONSTRUCTION DATA

# GENERAL CONSTRUCTION DATA

## The metric system

Linear

| | | |
|---|---|---|
| 1 centimetre (cm) | = | 10 millimetres (mm) |
| 1 decimetre (dm) | = | 10 centimetres (cm) |
| 1 metre (m) | = | 10 decimetres (dm) |
| 1 kilometre (km) | = | 1000 metres (m) |

Area

| | | |
|---|---|---|
| 100 sq millimetres | = | 1 sq centimetre |
| 100 sq centimetres | = | 1 sq decimetre |
| 100 sq decimetres | = | 1 sq metre |
| 1000 sq metres | = | 1 hectare |

Capacity

| | | |
|---|---|---|
| 1 millilitre (ml) | = | 1 cubic centimetre (cm3) |
| 1 centilitre (cl) | = | 10 millilitres (ml) |
| 1 decilitre (dl) | = | 10 centilitres (cl) |
| 1 litre (l) | = | 10 decilitres (dl) |

Weight

| | | |
|---|---|---|
| 1 centigram (cg) | = | 10 milligrams (mg) |
| 1 decigram (dg) | = | 10 centigrams (mcg) |
| 1 gram (g) | = | 10 decigrams (dg) |
| 1 decagram (dag) | = | 10 grams (g) |
| 1 hectogram (hg) | = | 10 decagrams (dag) |

## Conversion equivalents (imperial/metric)

Length

| | | |
|---|---|---|
| 1 inch | = | 25.4 mm |
| 1 foot | = | 304.8 mm |
| 1 yard | = | 914.4 mm |
| 1 yard | = | 0.9144 m |
| 1 mile | = | 1609.34 m |

Area

| | | |
|---|---|---|
| 1 sq inch | = | 645.16 sq mm |
| 1 sq ft | = | 0.092903 sq m |
| 1 sq yard | = | 0.8361 sq m |
| 1 acre | = | 4840 sq yards |
| 1 acre | = | 2.471 hectares |

Liquid

| | | |
|---|---|---|
| 1 lb water | = | 0.454 litres |
| 1 pint | = | 0.568 litres |
| 1 gallon | = | 4.546 litres |

Horse-power

| | | |
|---|---|---|
| 1 hp | = | 746 watts |
| 1 hp | = | 0.746 kW |
| 1 hp | = | 33,000 ft.lb/min |

Weight

| | | |
|---|---|---|
| 1 lb | = | 0.4536 kg |
| 1 cwt | = | 50.8 kg |
| 1 ton | = | 1016.1 kg |

**Conversion equivalents (metric/imperial)**

Length

| | | |
|---|---|---|
| 1 mm | = | 0.03937 inches |
| 1 centimetre | = | 0.3937 inches |
| 1 metre | = | 1.094 yards |
| 1 metre | = | 3.282 ft |
| 1 kilometre | = | 0.621373 miles |

Area

| | | |
|---|---|---|
| 1 sq millimetre | = | 0.00155 sq in |
| 1 sq metre | = | 10.764 sq ft |
| 1 sq metre | = | 1.196 sq yards |
| 1 acre | = | 4046.86 sq m |
| 1 hectare | = | 0.404686 acres |

Weight

| | | |
|---|---|---|
| 1 kg | = | 2.205 lbs |
| 1 kg | = | 0.01968 cwt |
| 1 kg | = | 0.000984 ton |

**Temperature equivalents**

In order to convert Fahrenheit to Celsius deduct 32 and multiply by 5/9. To convert Celsius to Fahrenheit multiply by 9/5 and add 32.

| Fahrenheit | Celsius |
|---|---|
| 230 | 110.0 |
| 220 | 104.4 |
| 210 | 98.9 |
| 200 | 93.3 |
| 190 | 87.8 |
| 180 | 82.2 |
| 170 | 76.7 |
| 160 | 71.1 |
| 150 | 65.6 |
| 140 | 60.0 |
| 130 | 54.4 |
| 120 | 48.9 |
| 110 | 43.3 |
| 100 | 37.8 |
| 90 | 32.2 |
| 80 | 26.7 |
| 70 | 21.1 |
| 60 | 15.6 |
| 50 | 10.0 |
| 40 | 4.4 |
| 30 | -1.1 |
| 20 | -6.7 |
| 10 | -12.2 |
| 0 | -17.8 |

## Areas and volumes

| Figure | Area | Perimeter |
|---|---|---|
| Rectangle | Length × breadth | Sum of sides |
| Triangle | Base × half of perpendicular height | Sum of sides |
| Quadrilateral | Sum of areas of contained triangles | Sum of sides |
| Trapezoidal | Sum of areas of contained triangles | Sum of sides |
| Trapezium | Half of sum of parallel sides × perpendicular height | Sum of sides |
| Parallelogram | Base × perpendicular height | Sum of sides |
| Regular polygon | Half sum of sides × half internal diameter | Sum of sides |
| Circle | pi × radius$^2$ | pi × diameter or pi × 2 × radius |

| Figure | Surface area | Volume |
|---|---|---|
| Cylinder | pi × 2 × radius$^2$ × length (curved surface only) | pi × radius$^2$ × length |
| Sphere | pi × diameter$^2$ | Diameter3 × 0.5236 |

| Weights of materials | kg/m2 |
|---|---|
| Lead sheeting, 2.24mm thick | 25.40 |
| Aluminium sheeting, 0.80mm thick | 2.20 |
| Copper sheeting, 0.80mm thick | 5.00 |
| Zinc sheeting, 4.60mm thick | 2.20 |

| **Weights of materials** | **kg/m2** |
|---|---|
| Profiled steel roof cladding | |
| 0.55mm thick | 5.30 |
| 0.70mm thick | 6.74 |
| 0.90mm thick | 8.67 |
| | **kg/10m2** |
| Roofing felt | |
| type 3B, glass fibre | 18.00 |
| type 3E, glass fibre | 28.00 |
| type 3G, glass fibre, perforated underlay | 18.00 |
| type 5B, polyester base, mineral surfaced | 38.00 |
| type 5E, polyester base, mineral surfaced | 38.00 |

**Tiling**

| | **Lap** | **Gauge mm** | **nr/m2** | **Battens m/m2** |
|---|---|---|---|---|
| Clay/concrete tiles | | | | |
| 267 × 165mm | 65 | 100 | 60.00 | 10.00 |
| | 65 | 98 | 64.00 | 10.50 |
| | 65 | 90 | 68.00 | 11.30 |
| 387 × 230mm | 75 | 300 | 16.00 | 3.20 |
| | 100 | 280 | 17.40 | 3.50 |
| 420 × 330mm | 75 | 340 | 16.00 | 3.20 |
| | 100 | 320 | 17.40 | 3.50 |
| Fibre cement slates | | | | |
| 500 x 250mm | 90 | 205 | 19.50 | 10.00 |
| | 80 | 210 | 19.10 | 10.50 |
| | 70 | 215 | 18.60 | 11.30 |

| | **Lap** | **Gauge mm** | **nr/m2** | **Battens m/m2** |
|---|---|---|---|---|
| 600 × 300mm | 105 | 250 | 13.60 | 4.04 |
| | 100 | 250 | 13.40 | 4.00 |
| | 90 | 255 | 13.10 | 3.92 |
| | 80 | 260 | 12.90 | 3.85 |
| | 70 | 263 | 12.70 | 3.77 |
| 400 × 200mm | 70 | 165 | 30.00 | 6.06 |
| | 75 | 162 | 30.90 | 6.17 |
| | 90 | 155 | 32.30 | 6.45 |
| 500 × 250mm | 70 | 215 | 18.60 | 4.65 |
| | 75 | 212 | 18.90 | 4.72 |
| | 90 | 205 | 19.50 | 4.88 |
| | 100 | 200 | 20.00 | 5.00 |
| | 110 | 195 | 20.50 | 5.13 |
| 600 × 300mm | 105 | 250 | 13.60 | 4.04 |
| | 100 | 250 | 13.40 | 4.00 |
| Natural slates | | | | |
| 405 × 205mm | 75 | 165 | 29.59 | 8.70 |
| 405 × 255mm | 75 | 165 | 23.75 | 6.06 |
| 405 × 305mm | 75 | 165 | 19.00 | 5.00 |
| 460 × 230mm | 75 | 195 | 23.00 | 6.00 |
| 460 × 255mm | 75 | 195 | 20.37 | 5.20 |
| 460 × 305mm | 75 | 195 | 17.00 | 5.00 |
| 510 × 255mm | 75 | 220 | 18.02 | 4.60 |
| 510 × 305mm | 75 | 220 | 15.00 | 4.00 |
| 560 × 280mm | 75 | 240 | 14.81 | 4.12 |
| 560 × 305mm | 75 | 240 | 14.00 | 4.00 |
| 610 × 305mm | 75 | 265 | 12.27 | 3.74 |

| | Lap | Gauge mm | nr/m2 | Battens m/m2 |
|---|---|---|---|---|
| Reconstructed stone slates | | | | |
| 380 × 250mm | 75 | 150 | 16.00 | 3.20 |
| | 100 | 140 | 17.40 | 3.50 |

| **Nails** | **per kg** |
|---|---|
| Steel round lost head nails | |
| 75 × 3.75mm | 160 |
| 65 × 3.35mm | 240 |
| 65 × 3.00mm | 270 |
| 60 × 3.35mm | 270 |
| 60 × 3.00mm | 330 |
| 50 × 3.00mm | 360 |
| 50 × 2.65mm | 420 |
| 40 × 2.36mm | 760 |
| Steel clout nails | |
| 100 × 4.50mm | 75 |
| 90 × 4.50mm | 85 |
| 75 × 3.75mm | 150 |
| 65 × 3.75mm | 180 |
| 50 × 3.75mm | 230 |
| 50 × 3.35mm | 290 |
| 50 × 3.00mm | 340 |
| 50 × 2.65mm | 430 |
| 50 × 2.65mm | 430 |
| 45 × 3.35mm | 330 |
| 45 × 2.65mm | 440 |
| 40 × 3.35mm | 350 |
| 40 × 2.65mm | 570 |
| 40 × 2.36mm | 700 |
| 30 × 3.00mm | 540 |
| 30 × 2.65mm | 660 |

| | Code | Colour | Thickness mm | kg/m2 |
|---|---|---|---|---|
| **Sheeting** | | | | |
| Lead | 3 | Green | 1.32 | 14.97 |
| | 4 | Blue | 1.80 | 20.41 |
| | 5 | Red | 2.24 | 25.40 |
| | 6 | Black | 2.65 | 30.05 |
| | 7 | White | 3.15 | 35.72 |
| | 8 | Orange | 3.55 | 40.26 |

| | SWG | Thickness mm | kg/m2 |
|---|---|---|---|
| Aluminium | 23 | 0.60 | 1.54 |
| | 21 | 0.80 | 2.05 |

| | Sheet size mm | SWG | Thickness mm | kg/m2 |
|---|---|---|---|---|
| Copper | 600 × 1800 | 26 | 0.45 | 4.04 |
| | 600 × 1800 | 24 | 0.55 | 4.94 |
| | 750 × 1800 | 22 | 0.70 | 6.29 |

| | SWG | Thickness mm | kg/m2 |
|---|---|---|---|
| Zinc | 9 | 0.43 | 3.10 |
| | 10 | 0.48 | 3.20 |
| | 11 | 0.56 | 3.80 |
| | 12 | 0.64 | 4.30 |
| | 13 | 0.71 | 4.8 |
| | 14 | 0.79 | 5.30 |
| | 15 | 0.91 | 6.20 |
| | 16 | 1.04 | 7.00 |

| **Roofing felt** | **Code** | **kg/10m2** |
| --- | --- | --- |
| type 1B, fibre | White | 18 |
| type 1E, fibre | White | 38 |
| type 2B, asbestos fine gravel | Green | 18 |
| type 2B, asbestos minrtal gravel | Green | 38 |
| type 3B, glass fibre | Red | 18 |
| type 3E, glass fibre | Red | 28 |
| type 3G, glass fibre, perforated underlay | Red | 32 |

# Index

Printed in the United States
by Baker & Taylor Publisher Services